Innovation, Technology, and Knowledge Management

Series Editor
Elias G. Carayannis, George Washington University, Washington D.C., USA

For further volumes:
http://www.springer.com/series/8124

Eskil Ullberg

Trade in Ideas

Performance and Behavioral Properties of Markets in Patents

Foreword by T. Hazlett

Springer

Eskil Ullberg
George Mason University
Arlington, VA, USA
eskil.ullberg@gmail.com
eullberg@gmu.edu

ISBN 978-1-4614-1271-7 e-ISBN 978-1-4614-1272-4
DOI 10.1007/978-1-4614-1272-4
Springer New York Dordrecht Heidelberg London

Library of Congress Control Number: 2011938481

Printed on acid-free paper

Springer is part of Springer Science+Business Media (www.springer.com)

To developing and developed country policy makers, economists, and business and university communities and individuals, interested in the role of the patent system in developing their economies by the exchange of human ideas

Foreword

Eskil Ullberg's book, *Trade in Ideas – Performance and Behavioral Properties of Markets in Patents*, concerns a problem of interest in economics since at least Adam Smith: how to extend the scope of the market? The market of particular interest is that for intellectual property, such as patents, or "IP." IP consists of technological know-how embedded in things, product, or production designs that are "new, useful, and nonobvious." By protecting ownership interests in such assets, Inventors will better realize the value of such innovation-creating inputs in ways that are more socially useful than the alternative.

That alternative was on display in Europe of the Middle Ages, prior to the advent of IP. Then, specialized techniques were beginning to be fairly widely applied, increasing productivity. But while the information on which these economic advances rested was a public good, and could have been applied to additional firms, industries, and markets around the world at essentially zero marginal cost, their developers sought to capture value by restricting their application. By the use of trade secrets, key knowledge was suppressed, and the value of advanced products or processes truncated. In the popular discourse of Pigouvian economics, value was maximized by private agents but social welfare was not.

Hence, the forces of economic evolution pushed for a more efficient legal regime. Driven by nation states that sought to stimulate local economies, policy makers – circa 1500 – began instituting ownership rules for technology. No longer would Inventors need to Invest so heavily in restricting knowledge of their efficiencies; they could share their productive assets and yet continue to reap the returns of ownership. Indeed, done well, those returns would increase from the payments of those additional entrepreneurs who were now able to more widely deploy them. What had been "free riding" would become profitable "gains from trade," the appropriators paying the Inventors. The result was economically powerful, the efficient sharing of productive resources whose marginal cost was zero.

This effort to expand the scope of innovation is traced by Eskil to the efforts of Italian and British governments pursuing direct foreign investment. Come to our land, bring us your inventions, make us more productive, and you will be paid a handsome fee. The bargain was straightforward: in exchange for revealing your profitable secret, the state will protect your ownership. The Inventor will expand the scope of an invention's deployment and save the high costs of secrecy. The public gains use of advances in business and science, and can itself use the revealed information to create still further improvements. The patent system is born.

Subsidies on Knowledge Production Versus Mechanisms to Sell IP to Diversify Risk and to Monetize Value

Fast forward 400 years. The institutions of property are now well established. Considerable investment is yielding economic growth, the 2% per capita per year wealth gains that economic historians have marveled at and which Joseph Schumpeter heralded as the defining achievement of capitalism. Now economists are curious about the IP system and how technical progress is created and then distributed across firms and markets. They see the public good problem as one that may not be solved by IP, but exacerbated. First, IP rights may be too weak to capture gains from basic research, knowledge breakthroughs that form essential inputs into profitable activities but cannot themselves be profitably developed and licensed. Another way to say this is that the fragmented, overlapping contributions of such inventiveness is – while collectively valuable to society as the "March of progress" – not subject to private appropriation of the parts. These contributions are too small and highly complementary and would, if owned individually, introduce more gridlock than the private incentives were worth – "tragedy of the anticommons" in modern parlance.

Second, once the basic advances are discovered, their public good nature suggests that the pricing schedule imposed by owners will inefficiently restrict output. Where marginal cost equals zero, so – for competitive equilibrium – should price. Yet price=0 yields no equilibrium, for it yields no product.

Noted economists such as Kenneth Arrow have tackled the problem and concluded: the market fails. Basic research is a public good, and it cannot be efficiently produced under the incentives of private ownership. Subsidies for its production should be instituted, and then public distribution – without price rationing the nonrivalrous good – should govern its deployment.

This, however, introduces a curious asymmetry. There is no denying that the use of the market – what Ronald Coase, since 1937, has been calling "the price system" – incurs costs. Property rights must be defined and enforced. Some trades will then occur; these transactions have costs of their own. Trades that occur will create gains, but some other potentially efficient trades will not due to the costs of using the price system. Prof. Arrow, in 1962, seized upon such imperfections and offered a solution put forth, in similar veins, by Prof. Hotelling in 1929, and Prof. Pigou in 1920. In one hand, market failure was asserted to involve a product created under conditions of declining average cost, an analogy to the IP production function; in the other hand, the market failure flowed from the presence of external costs or benefits.

Responding separately to each, Ronald Coase answered these critiques (in 1946, in 1959, and in 1960) with a burst of analytical clarity. He showed that the costs of the price system, while real, were not unique. Moreover, there was no cost-free alternative. Public institutions, including the tax and subsidy proposals put forth as remedies to "market failure," would themselves be offered under the constraints of "bounded rationality." There was no assurance that Government would know what investments to subsidize, what negative externalities to tax, or to what degree any of these public policy interventions should be performed. This led him to suggest that property rights, assigned to self-interested owners, would more reliably discover opportunities for gains than administrative agents operating without the incentives of wealth maximization and lacking the valuable information supplied by market prices. He could also have added the difficulties faced by public agents attempting to promote coordinated social solutions without having recourse to capital markets, institutions that, by offering to fund profitable investments, solve the challenging problems associated with the intertemporal matching of costs and benefits.

Prof. Arrow's argument that the state could improve upon markets due to the infirmities of the price system in producing socially useful knowledge was not implausible, but it was ill-formed. To assume positive transaction costs for one set of agents and not for the alternative regime was to commit, in Harold Demsetz' memorable phrase, the "Nirvana Fallacy."

Which is where Eskil Ullberg's timely and important research comes into the picture. He departs from the error made by Arrow, an ambitious leap, perhaps, but one that is in this case warranted. Eskil seeks to explain more of the mechanisms by which property rights, specifically IP, can be sold by Inventors to diversify risk and to monetize value. Using the methodology of experimental economics, he creates a controlled game in which players – rewarded with money returns, to the extent that they follow rules, manage risk, and execute smart trades – reveal how economic agents might generally transact in IP rights traded in organized exchanges.

The fundamental ideas motivating this research are two. First, the use of well-organized markets for the trading of IP rights presents certain theoretical advantages in advancing social efficiencies. Just as the earliest IP regimes tended to expand international trade and investment, commoditization of IP could potentially further extend cross-border flows by reducing the cost of accessing productive technologies. Specialization expands with the extent of the market; where Inventors can better capitalize their inventions, they will sink more resources in discovering more new, useful, and nonobvious products and methods. In testing how such trading institutions work, this research seeks to bring Adam Smith's trade theory up to the modern day work on market design by such scholars as Vernon Smith, David Porter, Steve Rassenti, and Charles Plott.

Second, in examining the manner in which IP rights are traded, Eskil pushes forward an empirical inquiry into the costs of using the price system. In a positive transaction costs world, there are no perfect solutions. But there are better and worse. This book seeks to reveal how trading might actually occur, and how the rules of the market can improve upon the operations of patent Inventors and patent Users.

How Is This Research Enterprise Conducted?

So how is this research enterprise conducted? It is both narrative and empirical. I have summarized and interpreted much of the narrative. I shall now describe, in thumbnail fashion, the empirical.

Laboratory experiments are conducted to observe how IP auctions can be constructed, how prices form, and then how prices change under alternative rules. The experiments are conducted as computer games played by groups of participants competing and cooperating, at the computer terminals, in real time. The primary set of experiments is labeled Experiment 1. Those playing play for real money, being paid a fixed fee of about $7 and then realizing about $30–$100 for about a 2-h session. Players are selected from a pool of George Mason University undergraduates who have been trained in such exercises and are used in multiple sessions. (There were 41 sessions run by Eskil all told, each lasting about 2 h or more.) This means that both the experiments benefited from the use of "expert" players and the results were observed wherein the players were learning strategies specific to these experiments.

Each experiment included three rounds wherein choices were made by three sets of players:

An Inventor (who has an IP right to sell)
One or two Traders (intermediaries who buy and then sell IP rights)
Four to Eight IP Users (firms that form the ultimate buyers of the patents) from two distinct industries, either of which places value on the use of the patent in question.

Round One is initiated by the Inventor, who posts supply (or offer) prices, for the asset she wishes to sell. That asset is, of course, a patent. The contract supplied, however, contains two features: a fixed price, plus a royalty. The former is a lump sum, one-time payment yielding ownership of the patent in the market occupied by the User. The latter commits the IP buyer to a stream of payments equal to the royalty rate times gross sales for some time limited period (i.e., the life of the patent). The price offers by the Inventor are informed by data supplied. Specifically, the general range of User valuations provides guidance as to demand. Relying on this, offer prices are posted.

Round One continues as demand-side bidders now join the auction. Traders and Users both register their bids. The latter reflect the valuations they are given, valuations that are uniformly distributed between high and low values and, hence, contain a stochastic element – introducing some level of risk. Users are from two distinct industries, and valuations systematically differ across the two (one is high and the other is low). When a User purchases a patent, however, it may only use it in its market. That leaves additional potential value – generated in the other market – untapped. Here is where the Trader has a profit opportunity. It can outbid Users to win the patent, and then split the patent into two rights – one serving Industry A, the other Industry B. Rights will then be sold to not one, but two, high bidders operating in distinct output markets. This secondary market activity also commences in Round One, allowing Traders to profit by splitting rights and selling to Users in either market. Because only Traders can split the rights, these intermediaries are necessary for the market to achieve maximum use of the patent (which is to say, deployment across both industries).

Round Two and Three is a quiet period for both the Seller and the Traders. The Users are now in their deployment mode, using patents purchased to produce new products (under the patent) or to block competition to existing outputs by owning the patent. In addition to using the patents, a secondary market allows Users to re-resell the contracts to other Users at the beginning of these rounds.

With the completion of three Rounds, payoffs to the individual players are registered. More importantly, of course, the experiment has produced data on the manner in which the patent rights are offered, demanded, and deployed.

What Eskil calls "linear pricing," and what I think of as "two-part tariffs," help provide an interesting twist. Contracts to assign IP rights include two payment components: a fixed fee and a royalty. This, along with provisions in the licensing mechanism and the market structure imposed in the experiment, allow for two different kinds of deployments: Blocking or Use. That is to say, a firm may purchase a patent in order to deploy the technology licensed, or to stop another firm from deploying the technology licensed. The latter could be undertaken to protect existing market share from dissipation. The significance of the payment scheme is that the fixed payment compensates the Inventor when the patent is used for blocking; the royalty compensates when the patent is productively utilized. Both can happen simultaneously, in different markets, when Traders outbid single Users and then split licenses.

In most comparative static models, actions taken to block competitive entry are per se inefficient. Eskil has a more dynamic set of efficiencies in mind. On the one hand, the Inventor receives the capitalized value of the patent, eliminating risks associated with variable future returns. The User, meanwhile, may better amortize the capital that it has invested, as it better knows – from existing sales and costs – the value of the blocking rights. This may reduce static efficiencies but by allowing investors to hedge bets, encourages dynamic gains via enhanced incentives for investment.

The basic result of this experiment is that, consistent with standard laboratory experience, prices converge fairly quickly to levels that produce gains from trade. Impersonal trade creates a market in IP. Yet, because the auction designer can calculate the total gains from trade available to players, it is possible to measure the extent to which the trading exploits all possible gains. On this score, Eskil cites somewhat disappointing results. Whereas a standard result is that Traders can typically, after sufficient learning, extract 90% of total gains (with one-part tariffs), Traders in this game realize only about 40% (with the two-part tariff).

Several features are varied in this basic auction template to observe how such rule changes induce distinct outcomes. The most important is that patent validity is altered from a baseline of essentially 100% validity (no patent enforcement issues) to just 38% validity. The latter yields a situation where the IP buyer may find that it has purchased an unenforceable patent. The winner who has paid a fixed fee is harmed by this; the winner who has paid a royalty has not. I note that this set of outcomes is highly dependent on the assumptions of the model, as real-world buyers of bad patents are most assuredly hurt when the complementary investments they make are then rendered less valuable.

Two main points can be gleaned from this research. First, it is efficient to separate the two aspects of the contract, the blocking deployment and the investment deployment. As Eskil writes, there are really, "Two products in one" (Chap. 1). But only if the patent is useful in two markets, and only if the rights are sold exclusively in each. These restrictions on both market structure and the form of the contract may be a good starting point for modeling the problem at hand. But there is a considerable distance between here and the actual markets we seek to understand.

The second point is that organized markets bring demand-side bidders closer to IP sellers, improving incentives for inventive activity (Chap. 2). Noting that Inventors seek to improve access to demanders, the book constructs mechanisms for bidders to reveal their values. In constructing a framework for IP to be commoditized and traded in impersonal markets, this research aims to further the process wherein valuable activities can capture more of the gains they generate, while shouldering less of the risks. The payoff in increased economic growth is straightforward.

Policy Hurdles to Tackle for Wide Spread Adoption of Mechanisms for Exchange in IP

Eskil is alert to note that "Organized markets [having these particular features] don't yet exist" (Chap. 1). The significance of this point is not lost. Owners of IP, not to mention a limitless supply of intermediaries, are capable of constructing the framework Eskil offers. While these actors own, trade, and deploy IP in some ways that are similar to the constructs in the paper, for example, the two-part tariff pricing structure, they do not generally enlist impersonal exchanges. The suggestion has been raised, and some efforts are indeed being made to change this. A law firm turned brokerage, Ocean Tomo is working on this problem in the USA. They want to bring liquidity to IP markets by publicly listing assets in a transparent and standardized format.

But the great majority of IP transactions are handled in other ways. Firms typically negotiate deals bilaterally. Many of the transactions turn into mergers or acquisitions, one way to eliminate double marginalization, e.g., effectively paying a fixed fee to the owner of the IP and then taking his entire company's IP portfolio. This was done, e.g., when Qualcomm purchased Flarion, owner of many OFDM patents in the wireless sector. (I thought it would be fun to mention Qualcomm, by the way, supposing that it would prompt the reader to think of another way the IP transactions are conducted – through years of litigation.) Patent pools and standard setting organizations provide other ways for IP owners to distribute their supply offers.

So far as organized markets can be advanced for IP, however, the enterprise is worth pursuing. Patents seem to present at least three issues that make these developments particularly challenging. First, commodities are difficult to standardize into transparent ownership rights. Common equity shares may seem equally difficult to standardize, but they are not. Shares are standard within the firm, and firms are organized units that produce business metrics – income statements and balance sheets, e.g. – that can be compared across firms. Indeed, organized exchanges, as well as the regulators that supervise them, actively impose accounting formats on publicly traded companies just so the disparate operations of the many listed firms can be evaluated. It is important to note that, when firms are in the start-up phase and can boast no such operating metrics, they are generally unwelcome on organized exchanges. Only when the company has matured to the point where its accounts can be compared to those of other firms will be listed and able to make its initial public offering.

Patents seem to be much more like venture stage investments than public companies. Their defining characteristics are idiosyncratic and expert knowledge is required to understand how the asset might be valued. And, speaking of Qualcomm, it should also be noted that IP is a very hot, growing area of legal practice. Enforcement costs for patents, in particular, are high and rising. Dynamic areas of the economy, including wireless networking and bio-tech, are virtual hotbeds of litigation activity. Three years ago, in a speech at George Mason University, the CEO of Qualcomm said that his wireless technology company was in danger of turning into a law firm. I suspect many local experts in the matter would heartily agree! This suggests that patents are not so easy to standardize and therefore not so easy to trade on organized markets.

This leads to the second serious hurdle. A given patent is generally just one input among many. The paper cites the impressive statistic that the average automobile produced today embeds some 2,000–3,000 patents. Combinatorial auctions can be used to value packages of assets, but the creation of institutions that can accommodate package bidding is not a trivial matter. Indeed, this effort is made difficult or impossible by the fact that there is a flow of new IP coming to market from disparate sources. The optimal combination of inputs is rarely sold as a package, but pieced together from both new and existing IP rights.

Third, and lastly, the public good nature of IP tends to favor widespread adoption. Inventors do not want the highest price bid, but the highest revenue. To sell the IP for its blocking value requires, however, the sale to be exclusive within the industry. (And patent resale, or secondary market licensing, must be forbidden or the blocking owner could undercut Inventor royalties in other markets.) Moreover it removes the simple metric of price as the relevant variable. Of course, Users (or Traders) can bid prices in the form of fixed fees, but then they shoulder all risks in developing and marketing the technology. The risk sharing associated with royalties has its advantages.

Eskil's laboratory testing of an impersonal exchange mechanism for IP rights finds patent auctions inefficient relative to other experimental auctions. This may be due to the design of his institutions; perhaps more efficient rules, strategies, and participant learning would improve the gains from trade. That surely forms an ambitious research agenda moving forward. But it may also be that Eskil's results are already a good reflection of reality.

Praise for Trade in Ideas

"This is a book for the times. Never have we been more in need of the wealth creation process that can only come from innovations subjected to the trial and error process of selection to decide what among all the experiments can be supported for further trial."

Vernon L. Smith, *Nobel Laureate in Economics 2002, Chapman University*

"Eskil Ullberg ... departs from the error made by Arrow, an ambitious leap, perhaps, but one that is in this case warranted. Eskil seeks to explain more of the mechanisms by which property rights, specifically IP, can be sold by inventors to diversify risk and to monetize value. Using the methodology of experimental economics, he creates a controlled game in which players – rewarded with money returns, to the extent that they follow rules, manage risk, and execute smart trades – reveal how economic agents might generally transact in IP rights traded in organized exchanges.

In testing how such trading institutions work, this research seeks to bring Adam Smith's trade theory up to the modern day work on market design by such scholars as Vernon Smith, David Porter, Steve Rassenti, and Charles Plott."

Thomas W. Hazlett, *Professor of Law & Economics, George Mason University, and Director of the Information Economy Project*

"Eskil Ullberg reports in this book perhaps the most rigorous experimental analyses yet to appear on patent trade. Anyone interested in market-based approaches to the protection of ideas will benefit from the insights found in this sophisticated work."

Dan Houser, *Professor of Economics, George Mason University*

Fairfax, VA Thomas W. Hazlett

Series Foreword

The Springer book series *Innovation, Technology, and Knowledge Management* was launched in March 2008 as a forum and intellectual, scholarly "podium" for global/local, transdisciplinary, transsectoral, public–private, and leading/"bleeding"-edge ideas, theories, and perspectives on these topics.

The book series is accompanied by the Springer *Journal of the Knowledge Economy*, which was launched in 2009 with the same editorial leadership.

The series showcases provocative views that diverge from the current "conventional wisdom," that are properly grounded in theory and practice, and that consider the concepts of *robust competitiveness*,[1] *sustainable entrepreneurship*,[2] and *democratic capitalism*,[3] central to its philosophy and objectives. More specifically, the aim of this series is to highlight emerging research and practice at the dynamic intersection of these fields, where individuals, organizations, industries, regions, and nations are harnessing creativity and invention to achieve and sustain growth.

Books that are part of the series explore the impact of innovation at the "macro" (economies, markets), "meso" (industries, firms), and "micro" levels

---

[1] We define *sustainable entrepreneurship* as the creation of viable, profitable, and scalable firms. Such firms engender the formation of self-replicating and mutually enhancing innovation networks and knowledge clusters (innovation ecosystems), leading toward robust competitiveness (E.G. Carayannis, *International Journal of Innovation and Regional Development* 1(3), 235–254, 2009).

[2] We understand *robust competitiveness* to be a state of economic being and becoming that avails systematic and defensible "unfair advantages" to the entities that are part of the economy. Such competitiveness is built on mutually complementary and reinforcing low-, medium-, and high-technology and public and private sector entities (government agencies, private firms, universities, and nongovernmental organizations) (E.G. Carayannis, *International Journal of Innovation and Regional Development* 1(3), 235–254, 2009).

[3] The concepts of *robust competitiveness* and *sustainable entrepreneurship* are pillars of a regime that we call "*democratic capitalism*" (as opposed to "popular or casino capitalism"), in which real opportunities for education and economic prosperity are available to all, especially – but not only – younger people. These are the direct derivative of a collection of top-down policies as well as bottom-up initiatives (including strong research and development policies and funding, but going beyond these to include the development of innovation networks and knowledge clusters across regions and sectors) (E.G. Carayannis and A. Kaloudis, *Japan Economic Currents*, p. 6–10 January 2009).

(teams, individuals), drawing from such related disciplines as finance, organizational psychology, research and development, science policy, information systems, and strategy, with the underlying theme that for innovation to be useful it must involve the sharing and application of knowledge.

Some of the key anchoring concepts of the series are outlined in the figure below and the definitions that follow (all definitions are from E.G. Carayannis and D.F.J. Campbell, *International Journal of Technology Management*, 46, 3–4, 2009).

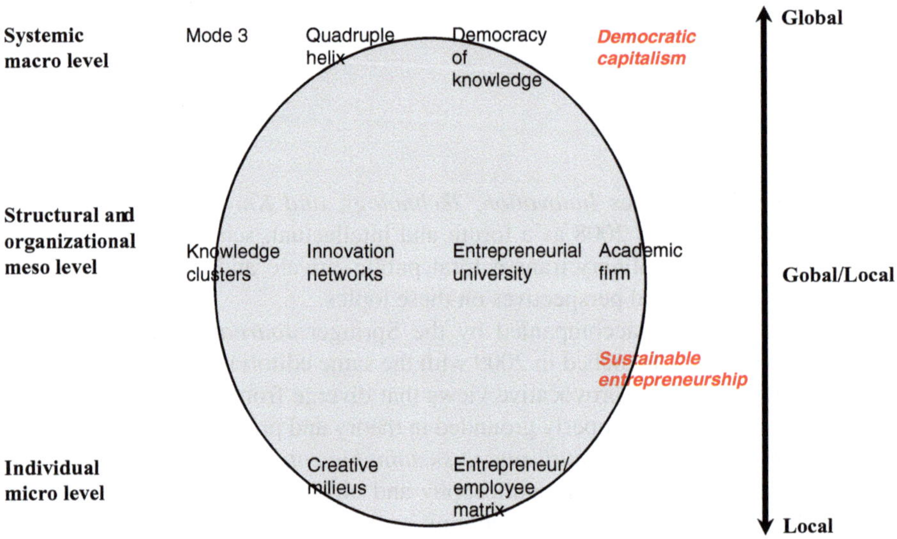

Conceptual profile of the series *Innovation, Technology, and Knowledge Management*

- The "Mode 3" Systems Approach for Knowledge Creation, Diffusion, and Use: "Mode 3" is a multilateral, multinodal, multimodal, and multilevel systems approach to the conceptualization, design, and management of real and virtual, "knowledge-stock" and "knowledge-flow," modalities that catalyze, accelerate, and support the creation, diffusion, sharing, absorption, and use of cospecialized knowledge assets. "Mode 3" is based on a system-theoretic perspective of socioeconomic, political, technological, and cultural trends and conditions that shape the coevolution of knowledge with the "knowledge-based and knowledge-driven, global/local economy and society."
- Quadruple Helix: Quadruple helix, in this context, means to add a "fourth helix" to the triple helix of government, university, and industry that we identify as the "media-based and culture-based public." This fourth helix associates with "media," "creative industries," "culture," "values," "life styles," "art," and perhaps also the notion of the "creative class."
- Innovation Networks: Innovation networks are real and virtual infrastructures and infratechnologies that serve to nurture creativity, trigger invention, and catalyze innovation in a public and/or private domain context (e.g., government–

university–industry public–private research and technology development cooperative partnerships).
- Knowledge Clusters: Knowledge clusters are agglomerations of cospecialized, mutually complementary, and reinforcing knowledge assets in the form of "knowledge stocks" and "knowledge flows" that exhibit self-organizing, learning-driven, dynamically adaptive competences and trends in the context of an open systems perspective.
- Twenty-First Century Innovation Ecosystem: A twenty-first century innovation ecosystem is a multilevel, multimodal, multinodal, and multiagent system of systems. The constituent systems consist of innovation metanetworks (networks of innovation networks and knowledge clusters) and knowledge metaclusters (clusters of innovation networks and knowledge clusters) as building blocks and organized in a self-referential or chaotic fractal knowledge and innovation architecture,[4] which in turn constitute agglomerations of human, social, intellectual, and financial capital stocks and flows as well as cultural and technological artifacts and modalities, continually coevolving, cospecializing, and cooperating. These innovation networks and knowledge clusters also form, reform, and dissolve within diverse institutional, political, technological, and socioeconomic domains, including government, university, industry, and nongovernmental organizations and involving information and communication technologies, biotechnologies, advanced materials, nanotechnologies, and next-generation energy technologies.

To whom is this book series directed? The book series addresses a diversity of audiences in different settings:

1. *Academic communities*: Academic communities worldwide represent a core group of readers. This follows from the theoretical/conceptual interest of the book series to influence academic discourses in the fields of knowledge, also carried by the claim of a certain saturation of academia with the current concepts and the postulate of a window of opportunity for new or at least additional concepts. Thus, it represents a key challenge for the series to exercise a certain impact on discourses in academia. In principle, all academic communities that are interested in knowledge (knowledge and innovation) could be tackled by the book series. The interdisciplinary (transdisciplinary) nature of the book series underscores that the scope of the book series is not limited a priori to a specific basket of disciplines. From a radical viewpoint, one could create the hypothesis that there is no discipline where knowledge is of no importance.
2. *Decision makers – private/academic entrepreneurs and public (governmental, subgovernmental) actors*: Two different groups of decision makers are being addressed simultaneously: (1) private entrepreneurs (firms, commercial firms, academic firms) and academic entrepreneurs (universities), interested in optimizing knowledge management and in developing heterogeneously composed knowledge-based research networks; and (2) public (governmental, subgovern-

---

[4]E.G. Carayannis, Strategic Management of Technological Learning, CRC Press 2000.

mental) actors that are interested in optimizing and further developing their policies and policy strategies that target knowledge and innovation. One purpose of public *knowledge and innovation policy* is to enhance the performance and competitiveness of advanced economies.

3. *Decision makers in general*: Decision makers are systematically being supplied with crucial information, for how to optimize knowledge-referring and knowledge-enhancing decision-making. The nature of this "crucial information" is conceptual as well as empirical (case-study-based). Empirical information highlights practical examples and points toward practical solutions (perhaps remedies), conceptual information offers the advantage of further-driving and further-carrying tools of understanding. Different groups of addressed decision makers could be decision makers in private firms and multinational corporations, responsible for the knowledge portfolio of companies; knowledge and knowledge management consultants; globalization experts, focusing on the internationalization of research and development, science and technology, and innovation; experts in university/business research networks; and political scientists, economists, and business professionals.
4. *Interested global readership*: Finally, the Springer book series addresses a whole global readership, composed of members who are generally interested in knowledge and innovation. The global readership could partially coincide with the communities as described above ("academic communities," "decision makers"), but could also refer to other constituencies and groups.

Elias G. Carayannis
Series Editor

Preface

This book is about trade in ideas. The special focus is the performance and behavioral properties of *markets* in patents with prices.

The framing of the problem comes from change in economic organization in history: from a *personal* exchange in technology (technical ideas) between people within hierarchies (guilds, firms, and nations) without legal rights to the *ideas*, to an *impersonal* exchange *between* hierarchies through markets based on the patent system, granting exclusive, and transferrable and licensable rights on technical ideas, making exchange with transparent (public) prices possible.

This shift began in Venice in 1474 with the creation of the first patent system. (Patents on technology surfaced a little earlier in Florence in 1421, as far as we know today, but the patent *institution* came half a century later). The system was put into law in the wake of financial difficulties of the dynamic city state, at the time known for its trade in industrially manufactured products and financial services, and was a deliberate economic policy to give incentives to *import* (trade), create (produce), and share (publish) the knowledge of new technology to boost *economic* development. The new policy replaced previous government subsidies and secrecy as mechanism for increased knowledge production. Honoring the Inventor's work would attract more people with creative minds to the city state, and create incentives to make previously held secrets public information, creating social gains through technology competition (learning by sharing), and market gains through specialization between Inventors, Investors (Traders), and Innovators.

These central issues of organized exchange, with prices, are at the heart of the study. It includes a patent licensing contract used to exchange the technology and mechanisms with different rules of trade. An informal discussion on what price (theoretically) aught to be paid for efficient outcomes – coinciding with what people do – decides much of the parameters. The study also implies for the role of the patent system as a trade system. A key patent parameter for tradability, the presumed validity (strength) of the granted patent, is included in the economic environment. The performance and behavioral properties of this complex institutional and economic environment, generating outcomes of allocation and prices is then studied in two laboratory experiments: The first focusing on prices and price formation and the second on coordination of investment decisions through price signaling. The experimental design institutions and environment allows testing the principles, capturing some key dynamic behavior of such exchange, making the case for parallelism to real-world trade in ideas. The experimental design is probably one of the first laboratory experiments in organized markets attempting to capture unique key characteristics of the patent system. Some of these characteristics may hopefully serve as inspiration for a broader scope of similar market design experiments.

The book begins with this brief economic history of the patent system. An experimental design is then addressed, introducing the components of the experimental environment and the institutions (the experimental economic *system*). The informal theory of the contract studied in the experiments is the proposed, suggesting that the *blocking value* of a patent (the value to "sit on" a right and use other technologies for now) is formally similar to insurance. The *investment value* (using the patented technology) is treated as a net present value investment under uncertainty (risk). The *trade value* is equal to "doubling" the market access through two noncompeting Innovators. This is the dynamic value. The mechanisms are then discussed in detail as the limited presumed validity of a patent (the chance it will be upheld in court if challenged). Each specialized hierarchy (Inventor, Trader/Investor, and Innovator) is given a definition by their action space. The economic system model attempts in a nut shell to summarize the basic principles of the patent system and how it has been and is used for the past 500 years. This analytical narrative has similarities in principle and to extant more personal exchange in patents.

The experimental sessions are then described in a historic narrative, one chapter for each experiment. This allows the reader to better understand the dynamics that are at play. A "participant study" is made asking the participants to describe how to succeed in the experiment to future participants. Finally, a discussion in statistical terms is done and policy conclusions drawn on trade in ideas based on the patent system.

Key results are that impersonal exchange in ideas with a fixed fee and royalty on sales contract supports the market access *insurance* hypothesis, and introducing rules of trade (institutions) doubles the performance of such markets. The typically very high efficiencies of experimental markets are not achieved. This may be a result of the experimental design or the risk preferences of participants, a behavioral issue not fully captured by the contract traded. These and other results are hopefully useful for policy development in a (uncertain) idea, or knowledge, economy, increasingly based on exchange in technical ideas through the patent system.

This book is a reduced version of my doctoral dissertation. I am grateful to a number of people who have made this process successful. The research is influenced by 20 years of management consulting in strategy and many years of working with the patent system strategy, especially from the European Patent Office. Here I would like to mention two people in particular: Ciaran McGinley, at the President's office, with whom I had the pleasure to work closely on strategic issues of the European Patent System for 6 years, and Dr. Ulrich Schatz, Principal Director of

International Affairs, who sponsored that joint work and introduced me to nearly every aspect of the global patent system at a level of principle and practice!

The academic research, whose findings this book report, has been carried out at the Interdisciplinary Center for Economic Science (ICES) at George Mason University, USA with affiliation to the doctoral program at the Royal Institute of Technology (KTH) in Sweden. Björn Hårsman, my principal adviser at KTH has given unwavering academic and practical support from the start. He also orchestrated much of the administrative and academic support needed for the project to come to a successful end. Critical funding of the project that made the research at ICES-GMU possible has been provided by the Savings Bank's Research Foundation (Sparbankernas Forskningsstiftelse), headed by Dr. Enrique Rodriquez. Åke E. Anderson has given invaluable advice on where to find excellence, and how to divide and focus academic work in relevant parts. I am also thankful to Stellan Lundström and Börje Johansson for their input on various parts of this manuscript.

Central to this research was understanding experimental economics, and I owe this understanding to the team at the ICES, especially Steve Rassenti, who invited me to come to learn experimental economics and do experiments and generously gave of his time. Many thanks to Vernon Smith for his sharp and constructive economic problem formulations, which launched the work forward into a working experimental design and avoided many dead-ends. Running several pilot-runs to observe the incentive structure prior to data collection turned out to be instrumental in arriving at a working design. David Porter gave insights in the problematization and economic system design which helped (together with Vernon's input and Steve's focus to look at personal exchange solutions from the real-world) to direct the research to solve the linear contract problem, the hardest nut in the whole research project. Dan Houser provided excellent input in discussions to solve the statistical analysis of these complex markets, making the statistic intuitive and understandable.

Finally, I would like to extend a special thank you to Springer's Nicholas Philipson and this book series editor, E. Carayannis, for making this publication possible, and to the accommodating editorial staff of Springer and SPi Global. In the process they challenged me to clarify my own thinking in several sections.

Arlington, VA Eskil Ullberg

Contents

1 From Personal to Impersonal Exchange in Ideas 1
- 1.1 Introduction 1
- 1.2 The Economic History of the Patent System: Productivity 4
- 1.3 The Economic Principles of the Patent System: Exchange 7
  - 1.3.1 A Market in Technology (Market Exchange) 7
  - 1.3.2 A Contract with the State (Social Exchange) 8
  - 1.3.3 The Economic Principle of Trade in Technology 10
- 1.4 The Economic Structure and the Patent System: Specialization 11
  - 1.4.1 Economic Organization and the Patent System 11
  - 1.4.2 Management of Risk and Uncertainty as a Principle of Organizational Limits 12
- 1.5 Impersonal Exchange Mechanisms for a Market in Patents with Prices: Coordination 14
  - 1.5.1 A Linear Contract 15
  - 1.5.2 A Proposed Informal Theory for Linear Contracts and Prices 17
  - 1.5.3 The Design Mechanisms for a Linear Contract 19
  - 1.5.4 Strength of the Patent System, Legal Uncertainty and Presumed Validity, in the Economic Environment 22
  - 1.5.5 The Role of Intermediaries, the “Dynamic Principle”, in the Economic Environment 23
  - 1.5.6 The Technology Space – Searchable Technologies – In the Economic Environment 24
- 1.6 Econometrics for the Experiments 25
- 1.7 Experimental Economics: Studying What People Do 26
  - 1.7.1 A Study of What People Do 26
  - 1.7.2 What Can We Learn from Experimental Data 30
- References 32

2 **A Dynamic Microeconomic System Design for Markets in Patents** 35
2.1 Introduction 35
2.2 Institutional Learning in History and the Two-Dimensional Nature of the Patent System 37
2.2.1 A Competitive Market in Technology 37
2.2.2 Allocation of Risk Bearing and Risk Sharing 39
2.2.3 Historic Development 39
2.3 The Dynamic Microeconomic System Design 41
2.3.1 The Legal Environment 43
2.3.2 The Economic Environment 44
2.3.3 The Institutions 50
2.4 Informal Theory of Prices and Initial Hypotheses 54
2.5 Summary 58
References 59

3 **Prices and Dynamic Gains in Markets for Patents: Experimental Study of Impersonal Exchange in Ideas** 63
3.1 Introduction 63
3.1.1 The Patent System as a Trade System 65
3.1.2 Summary of a Price System on Patents 66
3.1.3 The Mechanisms Design Criteria 68
3.2 The Mechanisms: Details of the Institutions, the Economic and Legal Environments 69
3.2.1 The Endogenous Experimental Flow 70
3.2.2 The Trading Screens 72
3.2.3 Subject Payments, Endowments, and Special "Rules of the Game" 74
3.3 Overview of Experimental Design, Design Parameters, Market Performance, and the Sequence of the Experiment 75
3.3.1 Market Performance and Dynamic Gains Under High Validity 79
3.3.2 Experiments with Low Validity 94
3.3.3 Secondary Market Reallocation 98
3.3.4 Homework 100
3.4 Linear Prices, System Gains, and Potential Capital Allocation: Hypotheses and Empirical Results 101
3.4.1 Linear Prices Compared with Proposed Theory (Static Analysis) 101
3.4.2 Comparison Between Institutions (Dynamic Efficiency) 107
3.4.3 Comparison Between Institutions (Heuristic Analysis) 108
3.4.4 Gains from Trade (Dynamic Analysis) 111
3.4.5 Capital Allocation (Calculation) 112

3.5 Summary and Conclusions: Prices and Dynamic Gains from Trade ... 113
3.5.1 Linear Prices and Informal Price Theory ... 113
3.5.2 Institutional Differences in Price Dynamics ... 115
3.5.3 Predictions of Prices ... 115
3.5.4 Gains from Trade ... 116
References ... 116

4 Coordinating Inventing and Innovating Through Markets in Patents with Prices: Experimental Study of Institutional Behavior and Price Signals ... 119
4.1 Introduction ... 119
4.2 The Trading System ... 121
4.3 Isolation of the Problem ... 122
4.4 The Economic System: Details of the Institutional, Economic, and Legal Environment ... 123
4.4.1 The Flow of the Experiment ... 124
4.4.2 The Trading Screens ... 126
4.4.3 Subject Payments, Endowments, and Special "Rules of the Game" ... 128
4.5 Experimental Design, Design Parameters, Performance, and the Sequence of the Experiment ... 129
4.5.1 Coordination and Gains from Trade Under High Validity ... 131
4.5.2 Coordination and Gains from Trade Under Low Validity ... 136
4.5.3 Secondary Market Usage ... 137
4.5.4 Summary ... 138
4.6 Coordination, Hypotheses, and Empirical Results ... 138
4.6.1 Does Convergence Take Place? ... 138
4.6.2 Differences in WTS Between Institutions and Validity ... 138
4.6.3 Coordination ... 140
4.7 Conclusions ... 140
References ... 141

5 Conclusions ... 143
5.1 Introduction ... 143
5.2 General Themes ... 143
5.2.1 Investigating Design Institutions Toward Impersonal Exchange in Ideas ... 143
5.2.2 How Risk is Shared in the Contract as Coordination Takes Place in a Market with Prices ... 144
5.2.3 Mechanism Design: The Integration of Information and Rules ... 144
5.2.4 A Dynamic Economic System ... 144

5.3 Experimental Results ... 145
5.3.1 Do Experiments Make Sense in a Complex Dynamic Environment Like the Patent System? ... 145
5.3.2 Why Are There No Organized Patent Markets Yet? ... 145
5.3.3 What Are the Main Theoretical and Experimental Results? ... 146
5.4 Further Research Propositions ... 149
References ... 150

6 Policy Propositions ... 151
6.1 Policy Propositions ... 151
6.2 Future Research to Clarify Further Policy Issues ... 156
6.2.1 Role of Patent System in Development ... 156
6.2.2 Insurance ... 157
6.2.3 Trust Between Firms ... 157
6.3 Concluding Remarks on Experimental Research ... 157
References ... 158

Epilogue ... 159

About the Author ... 165

Appendices ... 167

Bibliography ... 189

Author Index ... 193

Subject Index ... 195

Chapter 1
From Personal to Impersonal Exchange in Ideas

1.1 Introduction

Technology has always been at the heart of economic development. This knowledge has been kept private through a variety of means throughout history, but through the patent system, technology has become *tradable* in its own right. An excluding and tradable asset is created that has opened for a market in ideas, a process that has been going on for more than half a millennium.

The theme of this research is the transition of trade in technical knowledge and ideas from a *personal* exchange of know-how, trade secrets, etc., to an *impersonal* exchange of patents[1] in organized markets with transparent prices. This is thus a political economic problem. These types of studies typically encompass the legal system, market economics (institutions), and risk management implications. The transition has today reached a stage in which patents have become similar to physical assets with respect to validity, tradability, and geographic presence. This transition therefore possibly has an impact on general economic activity given the key importance technology has had historically on economic development. The more tradable intellectual property rights (like patents) create the foundation for efficiency gains which are at the heart of economic activity. Similar gains from specialization to those of industrial products and services could therefore be expected.

[1] Patents are tradable private property rights on technical ideas. Patents are granted by the state and give the holder excluding transferrable and licensable rights for *technical* ideas in exchange for disclosure of these ideas through publication. Recently in the USA, nontechnical ideas related to business processes have also been granted patents. Other intellectual property rights have similar properties for other ideas, such as copyrights, trademarks, etc. There are currently seven intellectual property rights granted under the WTO TRIPS agreement: copyright and related rights, trademarks, geographical indications, industrial designs, patents, layout designs of integrated circuits, protection of undisclosed information. (In addition, provisions for national control of anticompetitive practices in contractual licenses are granted, to stop practices that may have adverse effect on trade and impede transfer and dissemination of technology).

E. Ullberg, *Trade in Ideas: Performance and Behavioral Properties of Markets in Patents*, Innovation, Technology, and Knowledge Management 13,
DOI 10.1007/978-1-4614-1272-4_1, © Springer Science+Business Media, LLC 2012

The mechanism investigated is thus a *producer* market in patent-protected technical ideas. There are three values here: technology value (personal exchange), patent value (impersonal exchange), and consumer value (product and service exchange) of which the focus here is the patent value.

However, the standard literature on the economic analysis of patents has mostly been based on the *products* and expressed in terms of divergence between social and private gains (Arrow 1962; Plant 1934; Nordhaus 1969, 1972) or increased return from international trade (Krugman 1990; Schumpeter 1934). In this analysis, patent rights on technology are mostly treated as product monopolies. However, there is a lot of different technology, public and private as well as hundreds or thousands of patents from tens to hundreds of Inventors/owner that go into a *single* product or service today.

It is my contention that such an analysis, based on *monopoly* pricing of *products* (product value) – which is a *static* analysis based on monopoly theory assuming perfect information and has no institutions – may miss essential gains from a patent system coming from the *dynamic* exchange between specialized agents in a *competitive* trading of the *patent rights* themselves (technology value), and may lead to results that exclude trade aspects of technology altogether.

This experimental economic study was designed as a study of a contract (a tradable linear contract) on patent rights and impersonal exchange mechanisms used in trading the contract in an environment of specialized agents. The consequence of such impersonal exchange on the *dynamic* gains, performance and behavioral properties of the markets, and how the linear contract *prices* the patent values are then studied. In such experiments, the supply side does not know the "demand curve" (only a very broad value range is initially given the supply side), except through interactions in using the message space in the institution.

Two economic experiments are carried out; novel design market mechanisms for the linear contract are designed and carried out in a laboratory experiment with subjects. The first experiment investigates *prices* using different impersonal market designs for demand-side bidding on contracts on patents and the dynamic gains from such exchange, while the second experiment investigates *coordination* of allocation of resources for invention through transparent price signals for a defined technology "search area." The linear contract is introduced to trade the patents and an informal theory of prices on such contracts on patents is also developed and tested, using the experimental data. The study is thus characterized by a *dual* approach of a linear contract on patents traded in new institutional designs, studying market and economic system efficiency in an impersonal exchange environment.

This analysis differs from other neoclassical, game theoretical, and other experimental analyses in *pre*patent time like "patent races" or *post*patent time like "spillover" effects. Here, the analysis is entirely in *patent* time and exchange takes place through the institutions trading the linear contract. The patent system, which grants two rights (exclude and transfer/license), then provides the basis for a dynamic *trading system* for knowledge and ideas.

Before introducing the components and the real-world considerations put into the experimental design, I will here give the main results.

(a) The risks are *shifted* from the Inventor toward the Innovator (and ultimately the consumer) when competitive demand-side bidding is introduced. Such shift will probably result in a more competitive technology market and thereby technology-based economic growth. The experimental results show that buyers pay for the "blocking" value of the patent in the fixed fee, taking on risk and the "investment" value in the royalty (risk is *transferred* in the fixed fee and *shared* in the royalty of the linear contract). This risk-sharing differs between institutions but a fixed fee is always paid. The result also shows that experiments of the complex patent environment are possible and reflect key aspects of real-world behavior.[2]

(b) The *dynamic* market efficiency (as measured by the use of technology by Innovators) is increased by 100% when demand-side bidding is introduced, clearly indicating the dynamic value of the patent system. The total efficiency (about 40%) is not satisfactory for a real-world application (typically >90%) but that should thus be compared with about half of that when there is no, or little, demand-side bidding, a situation similar to today's personal exchange market. More mechanism design work and experiments – and probably some contract work – need to be investigated prior to real-world tests. The contention is that the patent system is a trade system.

(c) Coordination appears to work through price signals alone, using the linear contract. In terms of *willingness to search* broadly and coordinate to the highest technology values (most "useful" technology), demand-side bidding in one or both dimensions is important for gains to trade from coordination through markets with prices. The contention is that it is very difficult ("impossible") to dynamically efficiently allocate resources for invention and execute the allocation of a patent right under real-world conditions without a market. There are simply too many combinations. Krugman (1985) contends that supply side investments in technology (by governments) may lead to excess capacity and disappointment. The problem of comparative advantage between nations when it comes to technology (knowledge) may therefore be solved through the examined kinds of markets.

(d) The price theory is supported where a fixed price will be paid for "blocking" value that "insures" the *market access* of existing products. The contention is that a formal price theory based on insurance theory and investment theory – using the experimental data to make it more explicit – is the next logical step.

---

[2] The experimental study was carried out during 2006–2008, but recent events in the software industry (bidding on Nortel's patent portfolio in May/June 2011) illustrate that key aspects of the real-world dynamics are captured in the tested experimental system. In the Nortel case, the bidders evaluate blocking value, trade value (splitting the portfolio after purchase and license parts to others), and investment value in their own products or product portfolios, weighing their relative strengths against competitor's portfolios. Bidders on the Nortel portfolio-one billion dollar – in the end sold for 4.5 billion dollars – included Ericsson, Google, Microsoft Corp, Intel Corp, and RPX a patent-buying firm. These global companies and specialized patent holding firms thus apply the trade logic here presented in their market strategies based on technology portfolios in patents. See Kharif and Miller (2011).

(e) The demand-side bidding is generally more dynamic for the two-dimensional demand-side bidding institution. The contention is that the fixed fee and royalty prices can then *independently* express the blocking and investment values of the patents, making the linear contract critical for market efficiency.

The introduction of the components in the design generally follows an analytical narrative, except for the first section on patent system history which follows a historic narrative. Several readers of early versions of the text have encouraged that an explanation of the patent system should be included. The introduction is therefore extended but I hope it will meet that need of reference.

I will first give a brief economic history of the patent system, intending to show how it has been used for trade, followed by how a market in "patent products" has led to increasingly specialized agents (a new economic organization). The value of a patent and the contract and institutional designs – keys to increased market efficiency – are then discussed as elements of the impersonal market research. A short explanation of the experimental laboratory method then follows and how the performed experiments make use of this method to investigate studied patent markets which will end this introductory chapter.

1.2 The Economic History of the Patent System: Productivity

Intellectual property and technology in particular have always been at the center of economic development.[3] Such knowledge has been protected for private use through a variety of means of nondisclosure throughout history, such as guilds, trade secrets in firms, vertical integration, explicit nondisclosure agreements, joint ventures, mergers and acquisitions,[4] and classified military secrets. Guilds had private rules to pay for inventive activity within the community with royalties on their use to keep the art private within that "industry." As firms became organized (hierarchy), they

[3] Technological progress is considered by many sociologists, and perhaps also economic historians and technology historians, the single most important factor that has raised the standard of living for mankind. Medical advances only make a "dent" in welfare compared to advances in housing, sanitary, and agricultural technology. For example, Leslie White (energy technology), Gerhard Lenski (information technology), Alvin Toffler (service and information technology).

[4] A related point on mergers and acquisitions (M&A) was introduced by Coase and Demsetz who argued that transaction costs were a hindrance to M&A to control for harmful externalities. How property rights were construed then became the "make or break" question regarding market failure. In the case of patents, M&A has been the norm in the past for keeping technology private, in particular, in the Big Pharma industry. The construed rights argument has similarities with the patent system where the scope of the granted rights has been a constant matter of discussion for *centuries*: whether they are too broad, too narrow, not inventive enough, limits of subject matters, compulsory licensing, etc. It appears that currently, the rights tend to be narrow and low inventive step which is a situation not necessarily beneficial for market efficiency. It all seems to boil down to how we divide up the world and our ideas among us to the benefit of as many as possible.

began remuneration of in-house inventors. But through the patent system, technical[5] knowledge has become tradable, impersonally and in its own right.

The development of the patent system has taken place over a period of more than 500 years, beginning in Venice in 1474.[6] These initial "industrial" patents can be seen as issue to attract, i.e., *import*, technology for the benefit of economic development of the city, and later nation state. The value was in the manufacturing of the inventions. This appears to be a *political economic* motivation. Venice had lost trade (years before fall of Constantinople in 1453) and needed to renew economic development. These patents spread north and in the UK were often sold or given to "the friends of the king"[7] who traded them in exchange for market access to the UK for often *imported* manufacturing technologies. In 1623 in the UK, the crown's "industrial" monopolies were abolished after government abuse (using sale of monopolies to fund wars with France) and replaced by 14-year patents for product and process inventions, limiting the scope of protection from manufacturing areas to a much narrower product or process scope (de facto importing technology and exporting products). The parliament now controlled the issue of patents. This step probably introduced "blocking" as a strategy as competing technologies were developed. Another motivation across Europe was, according to some, to bring out technology from the guilds (who developed much of the technology in preindustrial Europe and much opposed the development of a patent system) and make the technical knowledge more useful to society by making it more public (disclosure) and keeping its ownership private at the same time (giving excluding rights to inventors). The UK patent law is also the model of US patent law. In 1836 in the USA,

[5] Today both technical and nontechnical ideas are patentable subject matter in the USA. The rest of the world grants, in principle, patents for technical ideas only. This is very contentious and currently there is a debate over what the scope of "technical" knowledge should be. Here we use technical knowledge for both types of ideas to represent the generally patentable ideas.

[6] This step was preceded by the guilds system and other ways to keep knowledge private. An even earlier step is claimed by Kaufer to be a change in view in society of *manual labor* during the early Middle Ages (520 AD) from "not deemed worthy of educated men" to "manual labor as cooperation with God in the task of creation." Kaufer continues: "In the Middle Ages, the pace of technological change accelerated dramatically. Italian cities like Florence, Lucca, Milan, and Venice became leaders in artisan production. However, secrecy, often enforced in the Italian city-states by draconian penalties, was used to protect technological advantages and avoid the disclosure of important know-how. This propensity changed with the emergence of a patent system." See Kaufer (1989). The recognition of the rights to *intellectual work* came 950 years after manual work was considered worthy of educated men in Europe through the patent law. '... And should it be legislated that the works and contrivances invented by them could *not be copied and made by others* so that they are deprived of their *honour*, men of such kind would exert their minds, invent and make things that would be of no small *utility and benefit to our State...*', '...infringer to pay him the sum of one hundred ducats and immediately destroy the contrivance...', '... But our Government will be free, at its complete discretion, to take and use for its needs any of the said contrivances and instruments, with this condition, however, that no one other than the authors shall operate them.' Italics added by author. This law, clearly motivated by economic policy, opened the way for *impersonal* exchange in technical knowledge.

[7] See, for example, Klee and Fairfield (1998).

examination was "invented" and patents were granted with presumed validity[8] and in 1982, also in the USA, the benefit of the doubt was given to the inventor, the agent taking the risks. This last step turned patents into assets much like physical assets with respect to legal validity and tradability, starting what is now commonly referred to as the "Pro-Patent era." On the international scene, the Paris convention from 1883 opened for international protection of patents and processes under non-discriminatory terms (national treatment) and giving the inventor a 1-year priority to file internationally (from date of invention/filing), expanding the *export* of patented technology, initially often accompanied by manufacturing ("foreign direct investments"), and in 1994 through WTO's TRIPS agreement, minimum standards for patent protection were set for member countries, including provisions for transfer and licensing of patents, creating provisions for a global exchange in technology.[9] In the end, all patents are still national or legal equivalents to national patents.[10] A central theme in this development can therefore be characterized as moving from a *personal* exchange (secrecy, guilds, joint ventures, etc.) toward an *impersonal* exchange of technical knowledge by means of developing more asset-like and internationally valid patents. The risk in such transactions is thereby reduced and incentives for trade are created. Efficiency gains can be expected from such impersonal exchange. See Fig. 1.1.

This history makes the patent system one of the more global economic systems today. Most recently there have been calls for a global patent system (harmonization) from industry and WIPO and for patent reforms in many countries, including the USA. However, competing systems may foster better economically viable patent systems. The general theme now seems to be adaptations for a more standardized way to deal with patents in a global economy.

---

[8] When patents are granted by the patent *office*, they are only *presumed* valid. They can always be challenged in courts and *deemed* valid (upheld) or invalid (invalidated). Until 1982, the burden of proof was on the patent holder when a patent was challenged, giving the benefit of the doubt to *society*, but after 1982 the benefit of the doubt was given to the *Inventor*, making the patent a much more asset-like *private* right.

[9] Regional and global harmonization attempts are the patent cooperation treaty, PCT, the international filing system run by the World Intellectual Property Organization, WIPO, in Geneva, and the European Patent Convention, EPC, the European filing and granting system run by European Patent Office and Organization, EPO. The PCT has recently adopted provisions making it one step closer to a global patenting system by amending Chap. II. PCT was originally thought of to be an "extension" of the Paris convention.

[10] There are increasingly calls for global harmonization of the patent system. This would create a "one size fits all" patent system, asking nations to give up national sovereignty over patentability to a world economic organization. In my view, I wonder if, at least for now, competing national/regional systems are not the best way to go. Competition means that certain economic qualities may develop better. This does not mean that the *procedure* has to be that different but that certain provisions are made for countries with national competitive advantage in certain areas. This is in any way how the system has been used historically. One can compare it with the international monetary system in this aspect. The trend, however, is clearly toward harmonization, especially since the TRIPS agreement. This was tried before 1883 but a system of principles of national treatment and priority year was put in place instead. Perhaps the time is ripe now?

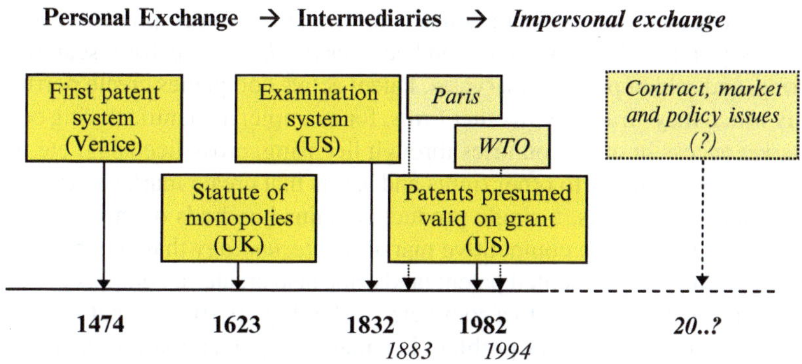

Fig. 1.1 From personal toward impersonal exchange in patents

1.3 The Economic Principles of the Patent System: Exchange

1.3.1 A Market in Technology (Market Exchange)

The patent systems of today typically give the holders two rights: the right to exclude and to transfer or license that right, creating the fundamentals for a market in ideas. These rights are thus transferrable for the holder and limiting action rights for others. In statute law, intellectual property rights, of which patents are one of seven (in TRIPS), are construed as a "chose in action" which differ from physical assets that are construed as a "chose in possession." However, Coase's view that physical property rights are really a bundle of limiting action rights ["…a right to perform certain (physical) actions." (Coase 1960), p. 44] suggests that there may be similarities in the economics between these two assets. The question studied here is thus whether these limiting action rights are beneficial to the economy as a whole from a trade economic perspective (not the trade-off between social and private product).

The excluding right also gives the patent *two* values: to exclusively use the technology in new, better products for a competitive advantage or to Block others' from using the competitive technology by "sitting on" the contract and instead using existing technologies of the firm. These values will be referred to as the *investment* value and the *blocking* value. Here, a dynamic is introduced that appears not to be captured in the neoclassical product monopoly (the competitive selection, allocation, and exchange process of technical ideas appears not to be included, at least not explicitly). These rights are today used in sophisticated and strategic ways. Research is a global activity since some time, manufacturing is almost global, and product and service markets are becoming global, protection is often sought where there is *competition* in research, manufacturing, or sale of products/services. The system is used to "protect" the coordination of activities performed in different nations, making use of national competitive advantages. This means that not only the main developed economies are critical for protection ("the West"), but also the developing economies where much manufacturing and also increasingly research is done and soon the least developed economies who begin to put an emphasis on education and property rights. The key issue then becomes *market access* for research, manufacturing, and sale of products/services. Patent active companies create portfolios of patents which they "trade" with other firms, for example, in manufacturing countries for market access in other countries through licensing, cross-licensing, etc. Patents are in this aspect similar to other rights and assets that create market access. Today, for all practical purposes, a single product contains hundreds or thousands of patents,[11] creating more of a competitive market for *technology* than anything close to a monopoly. Such a fragmented patent landscape also creates a complex combinatorial problem to obtain access to a complete "technology portfolio" that is practically useful in a product. The latter problem is a market organization problem. The lack of exchange mechanisms for these valuable and dispersed rights may also be a driving factor to try to develop mechanisms for exchange of quality patents.[12] A characteristic of this development is therefore the use of the patent system for strategically using or blocking competitive market access.

1.3.2 A Contract with the State (Social Exchange)

The patent holder still has to pay for the right like in the early days. First, inventors "pay" by disclosing the invention at a level such that a "man skilled in the art can reduce it to practice."[13] This informs everybody else about where the company is going with their products (it takes time between invention and production). This is true for all companies with a product life cycle of more than 18 months (the "silent period" between filing and publishing a patent application). This "social contract" is a key incentive to further research by "teaching the world" about the inventions. A more subtle side of this is that by seeking patent protection you are also disclosing what *problems* you are working on which may reveal strategic positioning.

However, it also informs competitors where *not* to put their research money and therefore directs the research in *other* fields, an alternative to "head on" technology competition. One company may therefore try one area. If commercially successful or not it will be known in due time. Others may then try other areas with outcome observed. There is thus a possibility for one company to intentionally try a new field for the benefit of others – there may be something out there for everyone[14] – hoping

[11] A car contains 2,000–3,000 patents from 200 to 300 *different* patent holders. *Source*: EPO.

[12] In the telecom business, where interoperability is a key, international standards are negotiated and when including patents (because patents are by definition state-of-the-art and therefore helpful in getting acceptance for the standards), patent holders must sign off their rights under RAND conditions (reasonable and nondiscriminatory terms). These kinds of problems are described, for example, by Heller in Gridlock economy. An answer may be combinatorial auction markets to recombine "too narrow" rights in useful blocks.

[13] Typical patent system criteria for disclosure of inventions.

[14] This is not to be confounded with a joint-venture type of research effort.

that others will reciprocate later, competing and cooperating at the same time. This may be seen as a *social exchange* among inventive companies made possible by the patent system. Actually such behavior was indicated in Experiment 2 by subjects reporting on the own behavior. See Chapter 4.

As mentioned in the previous section, making private technology publicly known was also a motivating factor behind the patent system. Today, the publication system of the patent office (after 18 months) allows for public information about patent applications, grants, and expired patents to be accessible by all. The "patent products" are therefore publicized by the state.[15] However, research indicates that about 50% of ideas are protected by trade secrets (keeping the ideas private) and roughly 30–35% are patented, making trade secrets still the most common way to protect ideas.[16]

Second, there is a fee to the patent office (the state), typically starting low and growing exponentially, allowing the inventor to pay if the invention is successful or drop it before any value has been realized.[17] This fee was originally given by "the friends of the king" to the king but now typically goes to the patent office (often ministry of justice or commerce) or directly to the treasury, often generating a revenue to the state. There is thus a financial incentive for the state to have quality patents, which are used for a long time. There is also an incentive for the inventor to reduce time to market (or the blocking right) not only for the limited time a patent is granted but also for the increasing renewal fees. The successful inventor then "gives back" some of the profits to the state.

These rights and obligations refer to the two parts of a patent: the disclosures (description of the invention) and the claims (list of granted excluding and transferrable rights[18]). The patent system can therefore be considered an *exchange system*

---

[15] A possible R&D agenda for companies with respect to patents is therefore (1) learning about technology through disclosures and staying out of infringements of claims; (2) using expired technology and getting new ideas; (3) developing new ideas into their own inventions; and (4) protecting and using them for products or trade.

[16] Reference to a European research report. There are many reasons for the trade secret use: patentability, enforceability, detectability, and secrecy to avoid revealing strategic research directions the company may take.

[17] In the US a fixed fee has been used but the fee structure is under change to a system more like the European exponential fees. These have the advantage that commercially viable patents are often renewed and other patents dropped. This has resulted in an *average economic life* of European patents of about 7 years (around 2005), far from the 20 years of *legal* life. The exponential pricing means "subsidy" of short lived/poor quality patents by longer-lived quality patents, an issue of interest for competition authorities.

[18] The scope of claims in one patent differs between systems. In the USA, many claims are allowed for one invention, whereas in Japan only one claim is allowed per patent. This makes USA and EU patents broader and, if upheld, powerful, whereas Japanese patents are weaker. The distinction is a result of an ongoing debate on the scope of a patent. Broad patents may lead to more risk-taking and investment in radically new ideas, whereas narrow patents may lead to smaller "inventive step" in the patents. See, for example, Nordhaus. A recent short study shows that the complexity of the applications is increasing in initial claims and total claims measured by issued claims. See Crouch (2008).

between inventors (information) and the state (money). This makes the patent system a complex policy issue and in a market analysis as in this case, the patent system becomes an agent (a legal environment).

1.3.3 The Economic Principle of Trade in Technology

The principles have thus remained more or less the same for the 500 years since its legal conception, but the Crown is now replaced by parliaments, broad "industrial" privileges replaced by narrower product, and process inventions given to individual inventors or assigned to firms (after some compensation) and "unlimited" or variable temporal rights, largely with a standardized time period of 20 years.

An important characteristic of the historic narrative – only briefly presented here – is that the development of the patent system appears to have been motivated by trade in technology. This motivation is different from the motivation referred to by most economists and economic textbooks of creating incentives for *inventions* in a closed economic environment (Arrow 1962; Plant 1934; Nordhaus 1969) without including national comparative advantages (Ricardo 1891, Heckscher 1919, Ohlin 1924), economies of scale (Krugman 1985), or other motivational factors of trade (e.g., wars). According to Plant, the patent system lacks theoretical economic principle (Plant 1934, p. 51), simply creating scarcity, but looking at the system as suggested here, as an exchange (trade) system, introduces the *economic principle* of trade and specialization as a motivational ground for the patent system. It is this economic principle applied to an economic environment including a patent system that is at the heart of the presented research.

In summary, the intention was originally to attract inventions made elsewhere and *import* them to Venice to further economic development there.[19] Later on, the international system (1883) allowed for export of patent-protected products, thus, *import* of technology for the receiving country, leading to economic development through, among other things, foreign direct investment. This appears to be the same mechanism at play as in Venice but at a product and process level. The international development may also be seen as an integration of markets for patented products. Some negative effects of this kind of trade have also developed. Since 1982, the rights have become more valuable as the burden of proof changed and infringements leading to litigation became costly. This has led to the emergence of a "patent market" in litigation. In this market, the judges set the prices. In order to avoid litigation, licensing agreements are increasingly made as out-of-court settlements under the threat of litigation costs (and/or uncertainty outcomes). Such settlements may "crowd out" smaller inventors as court costs literally explode.[20] This can be

[19] The first patents were related to water pumping, a German mining technology imported to Venice for use in mines on the main land controlled by Venice. See Kaufer (1989).

[20] The average court settlement is around 1–2 million dollars. Refer to patent law firms.

both a good and a bad development. So-called patent trolls amass patent portfolios secretly and then ask for licensing fees. The good part is that Users are more careful with what technology is used, but the inventors must be able to defend their newly acquired rights.[21]

The rights have thus become tradable and the licensing has allowed gains from specialization to emerge. One example is the university and (small) business patent collaboration (the Bayh-Dole act in the USA). Other examples are technologies such as IT and biotech where technology is often developed by smaller companies or specialized research centers and development by patent portfolio holding "marketing" companies. A broadening of patentable subject matter to nontechnology has also been made, primarily in the USA. One motive has been to give protection for new nontechnologies such as financial "technologies" and software. This is a very contentious issue and may well be reversed back to technology, the original patent system idea.

1.4 The Economic Structure and the Patent System: Specialization

1.4.1 Economic Organization and the Patent System

The development of the patent system as a trade system has given rise to specialized agents for different economic activities related to using technology, which is a *dynamic* economic change. The principle of personal exchange is here at work. The economic organization changes. This change has similarities to other areas of specialization where new markets are established. After the 1836 patent reform in the USA, the possibility of impersonal patent trade led to a rise in *specialized inventors* and market trade in patents, facilitated by *specialized patent attorneys* acting as intermediaries in this market (Lamoreaux and Sokoloff 1999; Lamoreaux and Sokoloff 2001). However, after some time, the integrated hierarchical model dominated. The reason why mergers and acquisitions dominated as an innovation strategy for the first 150 years is likely to be a problem of (1) transaction costs favoring M&A strategies; and (2) the way the patent system claims are organized, i.e., how the rights are construed, affecting the success of the underlying principle of tradability. It seems the patent policy focus has been on granting patents on technology and possibly less on the economic use of patents for trade. Such a shift to a "patent trade policy" may be economically efficient, facilitating the "trade value" of the patent system to be realized. Since the 1982 changes in the USA, when the validity and possibilities to litigate patents increased drastically, patent licensing has literally

[21] For this purpose, the EU has proposed creating a patent infringement insurance. However, there has not been any interest from the European insurance industry for these products to date.

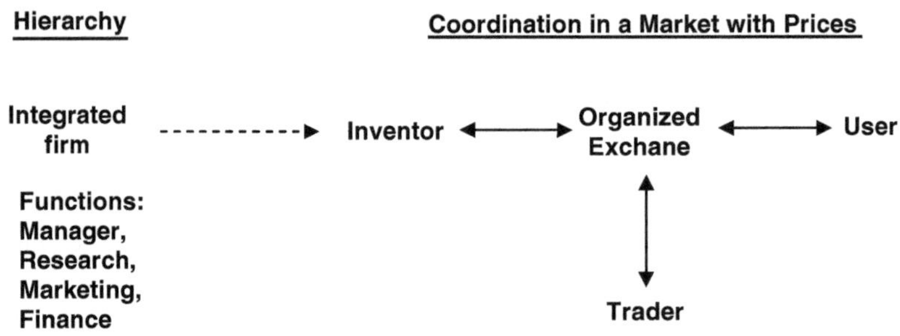

Fig. 1.2 Change in economic organization using a patent market

exploded and today accounts for up to 0.5 trillion dollar in the USA, and 1 trillion dollar in IP trade (including patents, copyrights, etc.), which is about 1% of the USA's total trade in products and services.[22] Similar developments can be seen in other countries. The specialization enabled by the patent system has gradually changed the business model from the *hierarchically* integrated firm as basis for economic organization *toward* a business model in which specialized agents coordinate their activities through a market ultimately with *prices*. See Fig. 1.2.

Such trade and specialization in patents described in the previous sections clearly give reason to believe that there are also essential gains from trade in technology made possible through the patent systems. The development of the patent system to make patents more tradable like physical assets is thus resulting in a *reorganization* of the economic activity at the core of the economy. Assuming the historic importance of technology to economic development, any changes in gains from trade in patents with respect to this "new" economic organization will therefore "multiply" into the whole economic system.[23] Several attempts in the last 10 years have been made to trade patents in more asset-like markets like patent auctions, but to date (2009) no organized markets with transparent prices similar to the impersonal trade of commodities in commodity markets and shares in financial market have emerged. I suggest that an important reason for this market failure is that it has shown difficult to find ways (rules) to create efficient trading.

1.4.2 Management of Risk and Uncertainty as a Principle of Organizational Limits

The standard view of economic organization of specialized firms takes its starting point in cost (Coase 1937). The approach suggested here differs in that the limits of

[22] See, for example, US census bureau on international trade and licenses and http://www.inventionstatistics.com on US patent licensing.

[23] Baumol, and others, contends that the greatness of the capitalist system lies in its incentive to create inventions, Baumol (2002).

the firm would be determined by the ability to manage risk in its customer relations, expressed as uncertainty in market access (sales and revenues), i.e., the starting point is in uncertainty in sales. Looking inside the hierarchical firm, we typically find a functional organization with market, finance, research, legal specialists, and general management functions. With tradable patent rights leading to specialization, these activities are increasingly separated into independent profit centers (previously cost centers) or more specialized companies and coordinated through internal transfer prices or external markets with prices.

My contention is that this division is motivated by the people (knowledge), systems (information and communication technology), and intellectual property rights (patents, trademarks, copyrights, etc.) needed to mange the *market access risks* related to these specialized activities. This would suggest that the change in economic organization due to the introduction of the patent system would be done according to the different level of risk they manage related to technical inventions.

Risk and uncertainty is here used with the following meaning: risks can be "put a number on," or "insurable" (Knight 1921; Schumpeter 1942; Barzel 1986; LeRoy and Singell 1987) and therefore managed through *markets* (in "insurance") but genuine uncertainty has to be managed through *people and systems* in a *hierarchy* (Ullberg et al. 2002).[24] This argument leads to the conclusion that risks that cannot be managed though a market (personal exchange with private prices or impersonal exchange with transparent prices) have to be managed in a hierarchical firm (personal exchange sometimes with transfer prices) defining the *limits* of an organization. Market exchanges have to be "sufficiently" efficient to make the separation advantageous for all parties. In other words, value has to be created through trade and specialization. The limiting factors are then on the one hand *personal* assessment of risk (to generate the "number") and on the other hand the existence of sufficiently efficient markets. A *system* with markets has to outperform a *system* without such markets.

In the case of patents and impersonal markets, what is managed can be divided into three clear activities: invention, finance/trade, and innovation. Invention here includes the cooperation of specialized patent lawyers. (The existence of the patent office is a fourth activity creating the fundament for such a market.) These activities are also observed in the real-world where inventor companies developing the technology are specializing from "market" companies, developing the products/services. Patent portfolio companies and other similar *intermediaries* using a financial business model of trading the patents represent yet another third key activity. These intermediaries are often keys to increasing efficiency in trade, since their incentives are to maximize each activity: in this case, inventing, using, and "multiplying" usage in different markets and products/services by limited license agreements.

The cost of capital of these companies is also typically very different, supporting the argument of risk being an organizing principle. Inventive firms typically have

[24] Reference to Knight (1921), Schumpeter (1942), Friedman (1976) on the subjective/objective probability and that the existence of a market in insurance could be a criteria to distinguish risk from uncertainty.

about 30% (at least), trading firms 5%, and innovation firms 10–15%. The allocation of risk-bearing is therefore also changing as a consequence of specialization. Gains from trade would lower the systemic risk. The extent of change in organization, and gains from trade that can be achieved, are therefore limited by the ability to manage risk through markets with prices (as in stock trade).[25]

At the heart of the development of economic organization from *hierarchy* to coordination in *markets*, is therefore the patent system's efficiency in *reducing risk* in the economic system, by trading technical knowledge between specialised agents and, the *dynamic efficiency* of the mechanism designs for such markets with transparent prices.

1.5 Impersonal Exchange Mechanisms for a Market in Patents with Prices: Coordination

In the previous sections, I showed how the patent system gives economic incentives for a shift from personal to impersonal exchange in technology through the dual rights of exclusion and transfer which increasingly leads to a new economic organization of specialized agents.[26] The coordination in a hierarchy (firm) is (gradually) replaced by coordination through a market (institution) with prices.[27]

In this section, I propose to present the key real-world considerations placed on the dynamic microeconomic system design used in the experiments to study this new institution. The actual model is presented in the summary section of the chapters, but this section gives some real-world background to the decisions made in designing the dynamic model. These considerations aim at creating a greater parallelism between the experimental outcome and hopefully real-world impact of policy proposals. The decisions focus on underlying *principles* of the legal, contractual, firm, and market rules used in the dynamic experimental economic model.

The focus on transparent prices and coordination makes the emphasis on factors important for increasing dynamic market efficiency. Since the *contract* used to trade and the *mechanism designs* (rules) are key factors in obtaining market efficiency (or contributing to market failure), these are discussed first. In this case, a perhaps more subtle aspect is also how the patent system itself is important, how the

---

[25] This is the classical contention of Adam Smith.

[26] Agents can here be independently owned firms or specialized functions – profit centers – within a company with the objective to actively trade the patent portfolio of a large corporation. These agents/functions perform the same function by providing material to sell in the market and have similar incentives to act "independently" of the owners producing business. In practice, the technology deemed not useful for the corporation is sold, other technology is used to strategically trade with competitors for market access (i.e., cross-licensing) and yet other technology is put in standards or simply never sold or licensed. In all these cases, there is a shift from the strictly hierarchical "trade secret" or product approach to a business in trading the rights themselves.

[27] This economic formulation of the problem was suggested by Vernon Smith.

rights are *construe* and *granted*, and the *validity* of patents. One could perhaps say that the contract between the patent holder and the state is *incomplete*[28] upon grant. The *intermediaries'* role (Traders) in creating the dynamics is also discussed briefly. I will first discuss the contract with a proposed price theory and mechanism design issues and then give the patent validity background, dynamic principle, and how technology choice is implemented.

1.5.1 A Linear Contract

To capture the dual values of the patent, the blocking value, and the investment value, I propose using a *linear contract.*[29] A linear contract can take on many forms but the one used in the study is the simplest form with an annual fix component and a royalty expressed as a percentage of revenues coming from the marginal value the technology adds to a product and the price it is sold for.[30] The reason for this deeper look at the contract at this stage is that this has been one of the most difficult issues to solve in the study – *what* to trade.[31]

1.5.1.1 The Linear Contract and Patents

There are two reasons for the choice of linear contract. First, the linear contract is a common contract that is used in today's *personal* exchange. It is difficult to get access to pricing information on patent contracts but one source of data on patent licensing contracts informs that 50% of contracts are linear, i.e., there is some fix component such as a lump sum payment combined with a royalty component on revenues (unit or dollar sales).[32] See Table 1.1. This suggests that in half of the cases there is an incentive to give a fix payment upfront and in others not. There appears however always to be a royalty (!). A lump sum payment can be seen as a "discounted" series of minimum or annual payments. They are only in 50% conditional on "signing" for the royalty part sometimes conditional (milestone). In other words,

[28] See the literature on "incomplete contracts," by Hart et al., i.e., Hart (1988).

[29] There is a rich literature on linear contracts. Basically, these contracts are contingency contracts on, for example, revenues.

[30] A similar procedure has recently been adopted by courts where a penalty has to be paid as a percentage on the value added to a product. Previously, the indemnification was on the price of the whole product.

[31] Special thanks to Prof. Smith for commenting that ordinary fix price contracts are "no different than potatoes and shoes," and do not express anything specific to the patent. Thanks also to Prof. Porter, who suggested that the problem might be a sharecropping problem. This led, via Stiglitz's work on incentives and risk-sharing in sharecropping, to the linear contract literature. Studying de facto patent licensing agreements also pointed toward a linear contract.

[32] Thanks to Royalty Patents Inc., in Washington, DC for providing this information.

Table 1.1 Types of payments in patent agreements

Royalty only	50%
Royalty only	29%
Tiered royalty rates	21%
Fix payment and Royalty	50%
Milestone Payment	12%
Lump sum payment	24%
Minimum Payment	13%

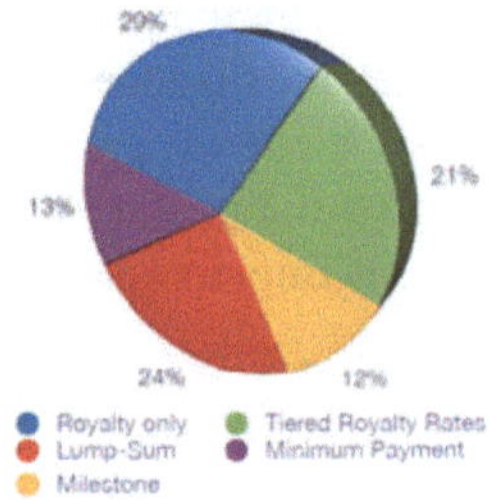

Source: Royalty Stat

risk-sharing is always present but risk transfer only in 50% of the cases and then sometimes conditional (milestone).

The motivating factor in this empirical case could be that the fixed fee acts as an incentive for the buyer to use (Invest in) the invention. It could thus help to deal with the *moral hazard* of selling the contract and leave the usage decision up to the buyer (principal-agent problem). Such decisions, which are typically project oriented in firms, may be given a low priority by the management.

The typical transition from personal to impersonal markets often follows, when it comes to the contract, the contract already used in the personal exchange.[33] In experimental economics, the way studies are done is also along these lines. A typical contract that is already traded, i.e., a *binding* contract, is concluded between buyers and sellers in the stock market-like double-auction mechanism. Then typically the efficiency of a mechanism design is studied.

1.5.1.2 The Linear Contract and Risk-Sharing

Typical linear contracts studied in the literature have followed similar functional characteristics, with input from fields other than patents. These contracts allow for a royalty to kick in at a certain level (minimum volume), tiered (different royalty rates given the usage/volume), flat fee, flat royalty, or a strict linear combination (minimum payment). It is the simple linear contract that is chosen in this study to capture both a fix (minimum) and a (constant) royalty payment. This type of linear contract thus allows for both risk *transfer* (the fix fee) and risk-sharing (the royalty). See Fig. 1.3.

With reference to Arrow (1962), a fix price market is used to *transfer* the risk, and separate derivate markets, one for each state of nature, are used to *share* the risks (Arrow-Debreu securities). The linear contract *combines,* in one contract, what the functioning of a fix price transfer market and derivate markets for each and every state of nature do to achieve optimal allocation of risk-sharing *and* risk-bearing

[33] The NY stock market (also commodities markets) is an example here. However, there is at some point an agreement, or standardization, upon what kind of contracts that can be traded. This agreement is an important factor in creating efficiency and transparency.

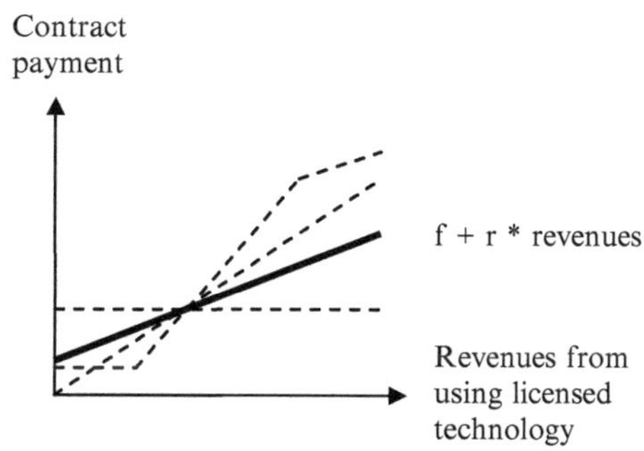

Fig. 1.3 Different types of linear contracts

in the economic system. In the proposed study, risk transfer and risk-sharing are thus negotiated in the same contract, reducing the number of markets needed to one and thereby increasing the liquidity since multiple risk preferences can be expressed in a linear contract price.

1.5.2 A Proposed Informal Theory for Linear Contracts and Prices

It is here proposed that the two *values* of a patent (blocking and investing) need to be *independently* expressed in order for optimal market efficiency in an impersonal exchange to be achieved. I will first shortly describe the difference in these values with respect to risk, then informally discuss a theory of pricing given these values, and finally what the optimal pricing of the contract would be, given the dual values.

The blocking value and investment values of a patent differ in terms of risk. The blocking value gives a holder a short-term protection against irreversible loss of *current* market access (sales) by "sitting on" the patented technology. (The firm's sales are then based on other technology the firm holds.) The firm can Block both others' products and own potential cannibalization. The current market access, i.e., the current sales of products and services that only have to be maintained, is relatively predictable in nature. Company analysts are regularly within a few percentages of actual sales. The investment value, on the other hand, gives a long-term protection of *future* market access (sales) based on technological *competition*. The future market access, i.e., future sales from future products and services which have to be designed, developed, market-tested, launched, upgraded, etc., is typically an entrepreneurial activity that is often highly uncertain in nature. It often takes years before a new product's market access can be accurately forecasted.

The blocking value is formally similar to *insurance*. See discussion on optimal insurance coverage (Arrow 1963, 1965; Smith 1968). Smith, expanding the tradition of Arrow for liability loss, makes the parallel between inventorying a product and liability insurance. An inventory of products "insures" the firm from "irreversible" losses of sales, just like liability insurance. In this case, the blocking patent right "insures" the firm from irreversible loss of sales. The way insurance is priced suggests that a proper price for this value would be a fix payment. One can therefore treat the short-term risk as an "insurable" risk.

The investment value is much closer to an investment calculation under uncertainty. Here, we borrow from investment theory and define such a value as the net present value of uncertain cash flows from the new product projects. Such risks are typically not transferrable (or "insurable") and have to be shared. A common way to share such risks is to agree on a royalty, which then shares the marginal benefit. What is left is the fix cost, which has to be discounted by the firm's project coast of capital. This risk therefore has to be managed in a hierarchy.

We therefore arrive at a *linear contract* with a fixed fee payment against the blocking value and a royalty payment on sales (value or units) against the investment value on the basis of the different risks of market access. The linear contract can thus allocate both risk-bearing and risk-sharing. A more elaborate discussion of the proposed price theory is done in a later chapter.

The next step is to define what an optimal price is. Here we use Coase's approach on optimal pricing from the marginal cost controversy (Coase 1946), where he outlines two principles for an optimal pricing system. The first should be that all buyers pay the same price irrespective of the contract (factor) usage. Otherwise buyers would not be able to choose rationally on the basis of price. The second is that the price should be fixed in such a way to allow allocation to the highest bidder. This implies that the amount paid should equal its value, which is, as Coase points out, equal to its cost of production. These criteria were outlined for fix price "contracts."

However, in this case, we are dealing with a linear contract with "two" prices: one fix price and one royalty price. One can see the patent as "two products in one" but the risk in the value of different "products," the usages, is such that a simple fix price is only optimal in one case and a royalty only in the other. The proposition above that the blocking and investment values are different in *nature* with respect to risk (existence of an insurance market and manageable only in a hierarchy) leads us to the proposition that to optimally price both these values, they have to be priced *independently*. This suggests that there also has to be *simultaneous* bidding for this contract so that bidders can state their (competing) preferences in the same market.

Using a simple game theoretical approach to bidding and bidding strategy, where a risk neutral agent bids up to its full value, the outcome is the following: If the high blocking value and high investment value bidder is the *same* agent, then the fix clearing price would equal to the *second* highest blocking value and the royalty price that yields a positive NPV of the *second* highest investment value given what is asked in the fix (second highest blocking value). If the agents having the highest blocking and investment value, respectively, are two different agents, then the NPV(Block, low risk) and NPV(Invest|Block, high risk) are compared. In this case, the contract will be given to the agent with the second highest blocking value. The investment value will be "destroyed" by the strategic blocker, outbidding a strategic investor (no royalty paid it not used). However, if the blocker decides to create a re-tradable contract (with acceptable royalty rates for investment which he does not know but can only find out through the bidding process), which he might do due to uncertainty in the future blocking value (as an exit strategy from the technology position) then the fix will be in *between* the second highest and highest blocking value and the royalty given by a positive NPV calculated using a value *between* the second highest and highest investment value given a fix *between* the second highest and highest blocking value, whichever is the highest NPV. The conclusion to this seems to be that in either case an Investor has to pay for a blocking value between the highest and second highest to get access to the contract. The royalty will be adjusted accordingly or be "destroyed" by a strategic blocker. In order to restrain the "blocker" from overbidding on the royalty, only a strong demand-side for investing may help, or high uncertainty in the technology competition. The latter is often the case in real-world environments.

In summary on the contract issue, my contention is that a tradable linear contract is necessary to express the different-in-nature values of a patent and a key factor to remedy market failure. Simply by transferring the patent in its own rights (with a fix price) may be inefficient.

The next section deals with the mechanisms used to impersonally exchange the linear contract, which are the key independent variable of the research.

1.5.3 The Design Mechanisms for a Linear Contract

The last part of the model is the heart of the study, the mechanism designs for dynamic impersonal exchanges of tradable linear contracts, and the use of experiments to generate data.

1.5.3.1 The Mechanism Design Criteria for Transition to Impersonal Exchange

There is a rich literature on fix price contracts both theoretically, i.e., Vickrey (1962), using game theory, i.e., Hurwicz (1960), and experimentally, i.e., Smith (1962). Also auctioning of linear contracts in *one* of the dimensions (typically the fix fee) has been studied. When it comes to patents, work has been done using neoclassical theory, possibly (Katz and Shapiro 1984), and game theory models pioneered by Kamien and Tauman (1986), but here also only for one-dimensional auction markets. When it comes to two-dimensional linear contract markets there is much less work (multipart prices by, for example, Coase constitute a similar result (two dimensions) but is based on different motivational grounds (fix and variable cost)).

Similarly in real-world cases of auctions today, a one-dimensional contract is used. Some real-world auctions have begun to appear and a range of companies, organizations, and specialized investment banks buy, sell, broker, license, cross-license, securitize, and thus also auction patents, much as after the 1836 US patent reform referred to earlier. These auctions may have started with bankruptcy auctions (in the USA) where patent rights are sold at a fix price like other assets. Other private initiatives use the same principles as that of auctioning art (similar to a Sotheby's art or car auction) to auction patents, also with a fix price. Government initiatives (tied to patent offices) are also using auctions to some extent for the purpose of increasing the use of technology, in particular presently un-used patented technology, by giving monetary incentives for technology transfer. In neither of these cases, it appears that the linear contract is the most common contract of patent licensing used.

The mechanism designs used in this study take their starting points in these one-dimensional markets but expand them to two dimensions to match the linear contracts. To be able to auction a two-dimensional contract is thus one criterion for the mechanism design.

A factor of importance to this study is also that the theoretical work often focuses on *supply side* bidding where sellers face a rational "price taking" buyer. However, the most common problem for inventors is getting access to buyers ("market access" for ideas) and to getting a price reflecting the patented ideas of economic value. Both these problems are of course solved by an efficient market institution design. A second criterion is therefore the emphasis on demand-side bidding.[34]

To make things more complicated, the linear contract is tradable and "valid" for three periods in the game, and can therefore be re-sold in periods 2 and 3 of the game. This take place in a secondary market where the contract is sold "as-is", and thus not negotiated. This transfer of the contract takes place using a fix price, i.e., a standard one-dimensional institution.

1.5.3.2 Experimental Economics Needed to Study a Dynamic System of Institutions and People

Furthermore, studying impersonal trade in patents when organized markets do not yet exist poses a data problem. To remedy this fact that information about impersonal exchange today is[35] private information, the *experimental economic* approach has been chosen as method of study. Experimental economics allows developing and studying a microeconomic system design which includes institutions of different mechanism designs and economic environment of contracts, agents, and technology, using human subjects. Experiments can also include rules for patent system dimensions such as multiple claims (fields-of-use) and presumed validity in a dynamic environment. An alternative to experiments is a game theory approach but the sheer complexity of such dynamics is prohibitive since it includes learning processes which currently cannot be modeled mathematically. A neoclassical approach would be even more challenging (no institution required for outcomes). In the next section, a more elaborate discussion on microeconomic systems as an

[34] An interesting result on experiments on monopolies yield competitive equilibrium prices when using a double-auction mechanism (Smith 1981).

[35] Data has been limited to personal exchange prices on patents and are hard to obtain (private information). Only a few patent auctions and a variety of internet based platforms have developed "listing sites" for potential buyers and sellers sometimes, most recently (2009) with bid/ask functions.

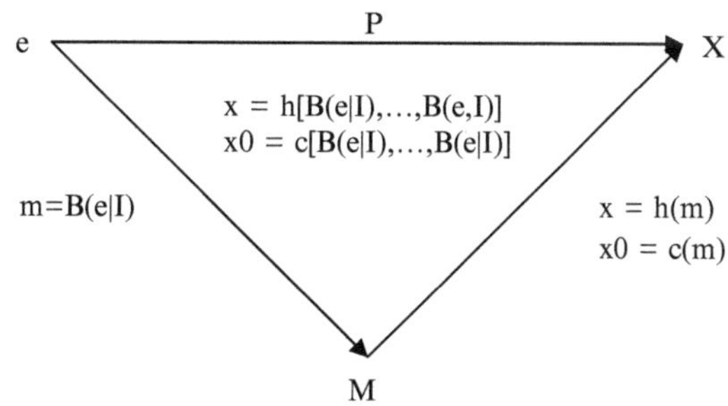

Fig. 1.4 A microeconomic system after Smith (1982)

experimental science (Smith 1982) and its relation to the conducted experiments has been done. I therefore propose mechanisms that can be implemented in an experimental design, dynamic, and accommodate the exchange of a linear contract under uncertainty.

1.5.3.3 The Designed Mechanisms

The way institutions are treated in experimental economics is here given a brief summary and then the implications for the study. In the (new) welfare economics (Reiter 1977), these institutions are treated as economic variables (Hurwicz 1973), whereas in experimental economics, as in this study, the institutions are treated as *treatment* variables, given different characteristics in experimental sessions.

A microeconomic economic system (Smith 1982) includes an *economic* environment (e) and an *institutional* environment (I), and results in allocations (h), payments (c), and outcomes (X) determined by agents choice *behavior* ($B(e,I)$) based on dispersed private information, given the institution. These three components constitute an experiment: environment, institution, and behavior. The institutions have a language of market (M), rules of communication and of contract, and a procedural structure. The language (M) consists of messages (m). This message exchange is started by a start rule, governed by a transition rule, and terminated with a stop rule, after which allocations (h) and payments (c) begin. These rules are managed by the institution through which the agents interact. The economic system analysis with institutions thus differs from the neoclassical approach in that the interaction between the agents given the incentives of the rules determines the outcome, not the environment and (rational) agent behavior alone, where buyers are price takers with a Pareto corresponding criterion (P). See Fig. 1.4.

The method of study here is to vary the language (message space) for demand-side bidding, keeping the supply side messages the same. This is the principle on which the study relies, together with learning from the well-studied fix price mechanism, to learn about the transition from personal to impersonal exchange in terms of market efficiency and dynamic system efficiency (including gains realized through Traders and "split" contracts). The sellers always ask a fixed fee (f) and a royalty (r). The buyers can then respond in three different ways, given the institution of study. First, the buyer can respond with an accept/reject of the standing ask, i.e., no willingness to pay can be expressed. Second, the buyer can respond with royalty bidding (the seller places a reservation value on the fix) in a double-auction fashion as in a fix price auction (the spread between "bid" and "ask" has to be reduced for each new bid). Third, the buyer can freely express willingness to pay in both fix and royalty by a double-auction procedure in both dimensions.

The first institution is most similar to today's personal, bilateral, exchange where inventors basically "knock on doors" for interest. The technical name for such an institution would be posted offer or two-dimensional manual Dutch clock auction. The second is similar to how linear contracts are auctioned today (in only one dimension with the other predetermined by the seller) but here the reservation value is on the fix and not the royalty, which is typical in those auctions. In the third, full expression of willingness to accept/pay, and thereby price discovery, can take place with independent bidding in *both* dimensions.

This design is flexible and allows studying several things. First, it allows for the study of how cash-motivated subjects actually *divide up* the blocking and investment values in the fix and royalty components, in particular, with respect to predictions by the informal theory (a nomothetic experiment). This is a central question of allocation of risk-*bearing* and risk-*sharing* through a single contract (without derivates). Second, it allows for the study of what *information/messages* are needed for efficient market outcomes using the linear contract. This is a question of information needed to trade patents (that have *two* values) (a heuristic experiment). Third, varying the demand-side messages also allows studying the limits of the theoretical predictions (a boundary experiment).

The resell marked is a standard double-auction mechanism with one price. Here, the buyer and seller can express their willingness to pay/accept using a single fix price bid.

In summary of the mechanism design issue, the purpose is to *jointly* study the linear contract and market mechanism for impersonal exchange, characterized by well-known equivalents (but not exact copies) for one-dimensional markets (to draw from that knowledge) and systematically increasing demand-side bidding space (to study the pricing, market efficiency, and allocation of resources for invention informed by transparent price signals alone) of a linear contract with only presumed validity. The resell market, open in periods 2 and 3 of the game, is a simple double-auction mechanism with a single price. Together these institutions form a dynamic experimental economic system both in "split" contracts (potential trader intervention) and time (primary, secondary market).

1.5.4 Strength of the Patent System, Legal Uncertainty and Presumed Validity, in the Economic Environment

An important concept used in the study is legal validity of patents. The patent system is made up of two bodies: the patent office, which grants the patent, and the patent courts (specialized national or federal, or part of the civil courts)[36], where patents can be challenged.

[36] It is not a criminal offense to "steal" an idea – "a chose in action" – but it is a criminal offense to steal a thing – "a chose in possession."

Patents are granted by people working in the patent office under criteria of newness, inventive step/nonobviousness, and industrial applicability/usefulness, which express the patent office's technical and legal opinion on patentability. Naturally, there is uncertainty in such a process, making a granted patent legally uncertain to some extent. Applicants have their own opinions, as do competitors, patents lawyers, and patent courts. The validity of such a patent right as it leaves the patent office is therefore a key dimension in efficiency. Patents are therefore only presumed valid upon grant by a patent office. The presumed validity then tells something about the "quality" of a patent.

The courts can deem a patent (in reality certain claims) valid or invalid. The presumption is based on an extensive examination of prior art in the technical area. However, there is still a chance that there is prior art that has been overlooked by the examiner or applicant (in some countries, the applicant is asked to supply known prior art for reasons such as later court hearings, speed up examination, etc.) and that the patent will not be held up in court if challenged. The examination process reduces the uncertainty with respect to validity. The holder can then confidently use the technology for sale or investment. In practice, the validity is about 38% in EU and was about 38% in the USA before 1982, measured by the percentage of patents upheld in court. After 1982, as the burden of proof changed from the infringed patent holder to the infringer, making it more challenging to overthrow a patent, the validity went up to 93%, i.e., practically as valid as a physical asset. Now the presumption of validity has two components, the actual presumed validity and the de facto validity of the legal claims including the cost of challenging and loosing a case.

Within the patent system, the examiner and courts have to balance the excluding right with social benefits (the social contract). Here, the incentives are challenging to say the least, since no one can possibly predict the full impact of an idea. Policies therefore tend to be either pro-granting or pro-nongranting. Some (applicants, examiners, businesses, judges, and politicians) believe that patents are good, while others think they are not so good or bad.[37] In this study, we simplify by stating that all these factors are in the "presumed validity," the probability that a patent is upheld in court if challenged.

If asked what the most important factor of a patent is, Users answer presumed validity of the right.[38] This "variable" is therefore of outmost importance for a well-functioning market. The presumption of validity is used in the microeconomic system design and experiment to introduce uncertainty in a contracted patent right and thereby of its value. Its value is varied between weak (38%) and strong (93%) to give information on the effect such a patent has on the questions studied (prices and coordination).

[37] Some even think that they are bad but since we have the system it would be worse not to have the system. Others think that it is the best of the worst. Mostly these arguments.

[38] See LES reports on annual investigation on licensing practices.

1.5.5 The Role of Intermediaries, the "Dynamic Principle", in the Economic Environment

The intermediaries play a role of maximizing the use of a patent right or collection of patent rights. They can buy a license of a patent from an inventor and then "split" this right into several limited rights for different technology markets, time, or field of use. They can also "bundle" groups of patent rights to cover commercially useful technologies. The field of use is used in this study to introduce this dynamics of the patent right in creating multiple market access for multiple products/services. The fields-of-use are defined by the legal claims.[39] The intermediary (Trader) has incentives to maximize the use of the technology as such. A producing company or industry is limited to their production system and inventors are even more limited and often have to stay very focused to generate sufficient returns. The Trader can, without limiting either agent's possibilities for market access, create "patent products" that are suitable for one or more industries. By doing this, they effectively increase the market access for the technology by allowing several noncompeting applications to use the same technology. This is a common practice among these intermediaries today, as they try to maximize value for their clients. In this study, it is a matter of patent system principle.

The reason why the Traders can do this profitably – although it is not that much easy – is that the market access risk of the license they buy to split is reduced by splitting into two "customer groups." More market access is potentially created, reducing the overall risk by "spreading" the risk over so many more consumers. To win a contract in the first place, they have to outbid the Users who wish to produce or Block. They can do this if they are successful in selling the two contracts afterwards. This risk-sharing is therefore an essential element of the dynamic gains from trading the patent rights and incorporated in the study at "Traders." By having a market mechanism with public prices, the role of the Trader can come in and create these dynamic efficiencies, *multiplying* the economic value. By doing so, they create a business model that can have a positive cash flow by taking a risk in the trading. To cover for that risk, capital is needed to back the trade (they do not have any human resources to invent or to produce). The agents may be seen as potentially defining a new role for capital:trade finance for ideas. They also create static efficiencies by adding to the number of buyers (for static portfolio trading). In Chapter 2, the economic incentives for this role are explored more in detail. But first what is traded, the contract, has be explained a little further.

1.5.6 The Technology Space – Searchable Technologies – In the Economic Environment

In the first experiment, the dynamics of pricing are investigated in an environment of Traders, linear contracts, and different mechanism designs (the first independent

[39] A patent application has two parts: disclosure and claims. The disclosure is the description of the invention. The claims define the excluding rights granted by the patent office and are for different "fields-of-use."

variable) under different patent validities (second independent variable). However, the "technology" (induced value set for contracts traded) is controlled by the experimenter. The second question of trade gains is related to the "signaling" of the linear prices and how the different institutional and legal environments affect the ability to allocate resources for invention to particular technical areas of different value. In the second experiment, the induced technology values are given a "search space" open to the subjects. Nine different "technologies" are randomly assigned to a "technology map" where subjects can choose among them. The only way to find out which is more valuable is to try an area and then observe the price, thus the price signaling. This second experiment thus includes a more flexible environment where (risky) coordination decisions can be observed through transparent prices studied in the first experiment. The values for this experiment are chosen such that there is only one technology focus with the highest value in NPV terms. However, the values span from high blocking value and low investment value to high investment value and low blocking value. Both the "search patterns," propensity to search and identify the high values,[40] can then be studied in the two contract dimensions for the different mechanisms and patent validity.

1.6 Econometrics for the Experiments

Econometrics for experiments is often nonparametric due to too few (and relatively costly) data points. In this case, a combination of standard parametric statistics and nonparametric tests has been used. However, in one case, comparing institutional performance, a special statistic has been developed due to the particularly challenging two-dimensional nature of the problem.

A fundamental problem in comparing institutions has been to turn the two-dimensional problem into a one-dimensional analysis. (Much effort was initially placed on finding a two-dimensional method, however, without success).

The price mechanism delivers prices expressed in fixed fee and royalty (f, r). A theoretical prediction proposed in the previous section suggests a certain combination of these prices with respect to the values (fix = blocking value, a royalty that satisfies NPV(investment value|blocking value) > 0). If stable price patterns appear in the experiments, then the different two-dimensional pricing behavior can be compared. (Stability in the behavior suggests that the microeconomic system has reached a Nash equilibrium, given the economic (contract and agents), legal (patent validity), and institutional environment (mechanism design). We can then conclude that the

[40] Using data from a European study of patent values Gambardella et al. (2005) to fit a distribution of technology values reveals that they appear to be distributed according to a *Poisson distribution*. This makes sense, as the Poisson distribution describes events of rare desired outcomes from many draws. This is often characteristic of inventions. A large number of experiments often has to be performed in order to have one useful outcome. And then even more experiments in the innovation process in the product and service markets. However, in this experiment, a uniform distribution is used for each technology to simplify calculations for subjects.

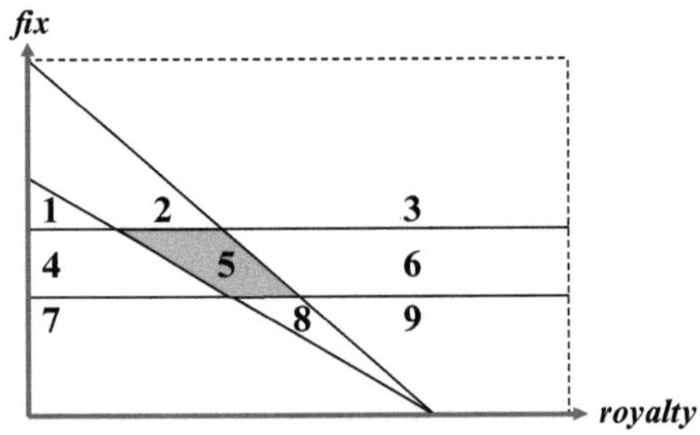

Fig. 1.5 "Price areas" or "boxes" are used in the statistical treatments. Area 5 indicates the predicted value

data can be the basis for such a comparison.) The solution to this problem has been to define nine "price areas" with the predicted price range in the middle (the values are uncertain and given as ranges) and higher and lower linear price combinations of (f, r) on the sides and above/below. See Fig 1.5.

Then, a frequency count of actual prices generated in the experiments that fall into these boxes informs us of the central tendency in prices given the institution and environments. The result is thus a frequency distribution of prices in *one* dimension, covering the two-dimensional prices. These *distributions* of boxed prices are then compared between treatments using a χ^2 test. The result is thus a statistical test of distributions. The null hypothesis is that the distributions are the same (we are unable to reject the null). The statistic tells us when we can reject this similarity and have indications of different price behavior between institutions. This leads to a way to pair wise (this test cannot rank the institutions) compare potentially different pricing behavior induced by the institutional mechanisms studied.

To test price predictions, a simple t-test is used and an F-test to *jointly* test the hypotheses for the fix and royalty.

1.7 Experimental Economics: Studying What People Do

Now, I propose to give a short background of experimental economics, the method of study, and how it is used in the two experiments. Readers accustomed with experiments can skip this section.

1.7.1 A Study of What People Do

Experimental economics was pioneered, developed, and used as a method by Nobel Laureate Vernon Smith in the 1950s and 1960s. The method is based on the study of what people *do* when presented with economic decisions. There are three components in this study: an economic *environment*, *institutions* through which exchange

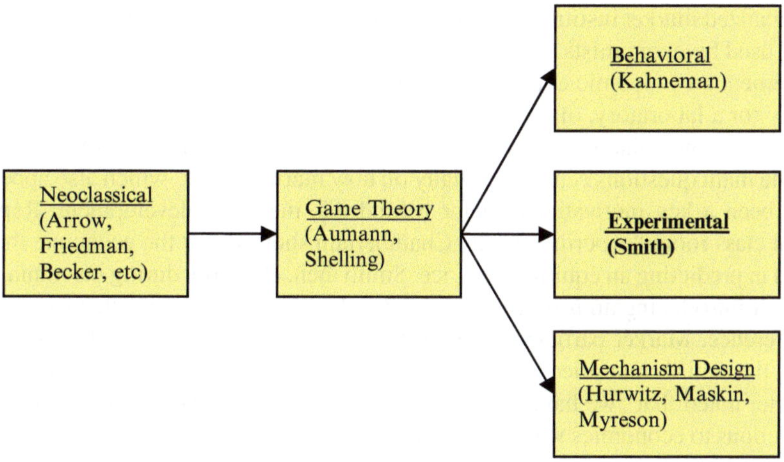

Fig. 1.6 Including *institutions by experiment and theory* and *behavior by experiments* in the economic analysis

takes place, and the subject's *behavior*. This is a very powerful approach since we cannot explain (theoretically) many of the key processes in how markets work – for example, the process by which people come to have common expectations and similar *learning* problems. However, such problems are solved every day by millions of people as they interact personally and impersonally, making literally billions of economic decisions. The abstractions that are used in neoclassical economics (rational expectations) and in game theory (common knowledge, rationality or bounded rationality) have led to the understanding of economic decision-making that we have today, but through experiments, we have been given the possibility of scientifically studying conditions under which these abstractions are supported and not supported by what people really do (Fig. 1.6).

One can classify various kinds of experiments by their methodological *objectives*. With experiments, one can establish the "laws" of *behavior* (nomothetic experiments). In these experiments, theories of behavior are tested. However, experiments are not limited to areas where there are mathematical models available to describe market interaction. Instead, experiments can provide data to *inform* new theories (heuristic experiments). Finally, as theories are validated (or falsified) using nomothetic experiments, experiments can also be used as a "robustness test" to test the *limitations* of a theory (boundary experiments). This classification was suggested for economic experiments by Smith in (1982) and draws from experimental scientific methodology writings by philosophers of science (Popper 1959; Hanson and Humphreys 1970; Hanson 1971; Kaplan 1964). Other classifications, by *function*, are also suggested by Smith, where combinations of environments (e) and institutions (I) make up a universe of $e \times I$ experiments. The *outcomes* or *messages* (bidding strategy, etc.) of the exchange are then compared with theory – for example, oligopoly theory, comparison of different auction mechanisms and rational expectations theory. The experimental toolbox is thus a very rich tool for economic analysis and advancement of the understanding of human behavior, in particular, with respect to organized market institutions (impersonal exchange), and has basically just begun to be used by economists. One reason for this is that it is challenging to design and implement an economic experiment. Another reason is that it is expensive; one has to pay for a laboratory, often extensive software development, and the subjects. Yet another reason is that there is a great inertia in how universities generally operate.[41]

The main questions centered initially on how markets work, which also appears to have been a key motivational factor behind this method's development. Repeated initial class room "experiments" by Chamberlain showed that the economic theories failed in predicting an equilibrium price. Smith then, allegedly during insomnia, wondered if introducing an *institution* where the clearing prices were listed would make a difference. Market participants could then learn about the prices. To his big surprise, the equilibrium price was found rapidly and experimental economics was born. He later noted that "we haven't gotten over it [the surprise] yet!" The importance of institutions to economics was also developed in parallel, initially without each other's knowledge, by game theorists led by Hurwitz.

The most important conclusion from 50 years of research is probably that market equilibrium prices converge with much fewer agents (commonly thought to be 3–6) than predicted by the theory (a large number so that everyone buyer becomes a price taker). Such a result suggests that there is a *learning* process taking place in the message exchange that is very effective. In an organized market such a *social* exchange takes place through impersonal exchange of messages given the rules of the institution. The method therefore spans from personal exchange to impersonal exchange, a fact that is used in this study.

The method is distinctly different from behavioral economics in that experimental economics studies what people do when they interact through *institutions* with rules. Since most economic transactions take place through institutions with rules, formal or informal (posted offer in the super market, haggling in the "traditional" market, capital markets, money market, etc.), experimental economics allows studying economic decision-making in a microeconomic *system* consisting of environment, institutions, and human subjects. The study of efficiency of institutions – realized outcome compared to theoretical optimal social outcome – is then made possible in a controlled laboratory where the institutional mechanism design (as well as the environment) can be an independent treatment variable.

Subjects are here not rational price takers with full information but a *social* exchange takes place between the subjects, through the trading *rules*. The experimenter tries to control for all other factors. In particular, and crucial to experiments, subjects are paid according to how they do in the experiment (performance payment) to induce profit-seeking behavior. This reward has to be "salient" to dominate any other motivational ground like fun to be in an experiment, destroying the experiment, etc. This point is worth mentioning since in some countries, paying subjects anything but a flat fee is not allowed. They are considered employees under contract.[42]

[41] According to some, a delay of 50 years between innovation and main stream application is common.

[42] This shows that the work done by subjects is pretty close to real-life working situations, at least when it comes to taxation.

Subjects are typically undergraduate students at universities. Some have argued that these are not representative in behavior. However, studies comparing them with professionals in, for example, finance show that students with some training do better in these experiments than professionals. The assumption is that the professionals come with their world view of how, for example, their market works, and act accordingly. The counterargument is that the young students will in fact soon be in business and people do not change that much in their preferences over life. The most fundamental argument, however, is that what is studied in experiments is inherently human behavior ("what people do"), not professional's or certain group's behavior. We have also learned that in experiments, as in much of real-life, "everything matters."[43] This makes controlled experiments challenging to do. A lot of (mostly undocumented) experience is needed, just as in other experimental sciences. In a sense, that is the price to pay for the flexibility experiments given to the economist.

Experimental economics investigates microeconomic *systems*. The systemic approach increases the parallelism with real-world dynamic economic decision-making (compared with neoclassical theory without institutions), allowing policy related problems to be studied with greater impact from what people would do in these systems given what other real people would do. A formalization of the experimental approach follows the lines of *static* equilibrium where a microeconomic system (S) is constituted by an economic (e) and an institutional (I) environment, $S=(e,I)$. The economic environment is formalized as having economic agents, commodities to trade including resources and agent's characteristics such as utility function, technology (knowledge), and commodities endowments. Often, technology is endowed at the beginning of the experiment but technology can also be included as a *learning process* (or search process) by specifying the limitations of search in the initial environment. The latter aspect is used in Experiment 2, where technology search is included in the dynamic environment, informed by transparent prices from the market designs tested. See Smith (1982) for a formalization of the microeconomic system.

However, in this experimental study, the "patent system" is included as part of a legal environment (L). The system is thus expended to include an economic, legal, and institutional environment. The legal environment is the patent validity and is thus a *treatment* variable in the experiments, together with the institutions. The system can then be formalized as $S=(E,I)$, where $E=(e,L)$. The experiment is based on a dynamic experimental model containing two markets: a primary market where the linear contracts are initially priced and a secondary market where these contracts can be re-traded (for a fixed fee). This dynamic experimental economic model can be seen as made up of two static systems as described here.

A parallel theoretical track to experimental economics has been *mechanism design* based on game theory. This work was pioneered by Hurwicz, who shared the Nobel Price in economics in 2007 with Maskin and Myerson. In this work, a formal structure for the design and evaluation of allocation mechanisms is expressed as economic variables. Game theory thus introduces institutions in the allocation process just as experimental

[43] A famous example here is a dictator game experiment where *one* word was changed in the instructions with the effect on the outcomes, "opponent" was changed to "partner." This apparently changed the framing of the game for the participants and taught experimenters that "every word matters."

economists do (as treatment variables). The mechanism design theory therefore allows for designing a better system than what is in place today, i.e., what can be observed today. It is thus a dynamic economic analysis where the structure of the economic system is to be regarded as an unknown. Mechanism design allows designing new mechanisms and exploring the constraints and tradeoffs. To assess one's efforts in the design, the work can be tested for *incentive compatibility*. The choice here is if the behavioral patterns constitute a Nash equilibrium, i.e., if no agent finds it advantageous to depart from his behavior pattern so long as the other does not. The proper integration of information (messages) and incentives (as given by rules) of resource allocation models (institutions) is still unsolved in mechanism design. See Hurwicz (1973). This integration is a key issue in the investigation (information/messages and rules) and is explored in three mechanism designs in the experiments. One could argue that there is a certain element of constructivism to these two economic disciplines.

An adjacent branch of economics, *behavioral economics,* studies human decision-making focusing on economic decision-making *without* an institutional body (thus people interact without trade rules). A recent key contribution is *prospect theory* by Kahneman and Tversky, which integrated psychological research into human decision-making under uncertainty. Kahneman shared the Nobel price in 2002 with Smith. Prospect theory suggests that people express loss aversion (Kahneman and Tversky 1979). As will be shown, loss aversion turned out to play an important role in the experimental design and had to be taken into account (or used) to get a functioning experiment.

One can see these developments as three steps toward an economic analysis including more of the human behavior: neoclassical, game theory, and mechanism design more or less in parallel with behavioral/experimental economics. Experimental economics stands out by making a clear distinction between the effect of the economic environment and the effect of institutions through which people interact and that few limitations are placed on people's behavior (by using real, cash-motivated, people). The price to pay for this flexibility in the economic analysis is the development of an experimental *economic system design* which is still largely considered an "art" (there are few textbooks on "how to" on all the intricate design issues involved). The design process often (always?) involves testing and development with subjects, to achieve a functional design for the subject tested. Developing an experiment on markets is therefore also a learning process for the experimenter on how (simplified) markets really work and how to develop a design that tests the particular aspects of interest. In this sense, economic experiments are very similar to experiments in nuclear physics at CERN, but with less costly equipment needed. What was characteristic with these experiments in physics was that, at least until the 1980s, one theory was needed for each experiment and few general models were developed.

1.7.2 What Can We Learn from Experimental Data

The central feature in experimental market studies is that the messages and rules that are used in the interaction (information in bids and asks) affect the incentives of

Table 1.2 Experimental economics and the experiments in this research

Why economists conduct experiments?	With reference to Experiment 1: Prices	With reference to Experiment 2: Coordination
1. Test a theory, or discriminate between theories	Test of proposed informal price theory for trading linear contracts on patents (testing risk transfer/sharing aspects)	Testing proposition that demand-side bidding can yield coordination of inventive, innovative, and trading activities
2. Explore the causes of a theory's failure	Test of pricing behavior for "boundary conditions" of blocking and investment values	
3. Establish empirical regularities as a basis for new theory	Documenting pricing behavior under different induced values, demand-side competitiveness, etc. for formalization of proposed informal theory	Investigating signaling properties of market designs on allocating resources for invention
4. Compare environments	*Key variable*. Robustness test to stress the informal theory by (1) varying presumed patent validity and (2) randomly varying induced blocking and investment values	Key variable
5. Compare institutions	*Key variable*. Three design market mechanisms were developed and used, varying the bidding language (message space) for demand-side bidding. Two, one, and zero dimensions of the fix and royalty prices in the linear contract were tested	Key variable
6. Evaluate policy proposals	Some work on patent "trade" policy. Market and system efficiency was compared between an institution resembling today's personal exchange and two design institutions that were impersonal	
7. The laboratory as a testing ground for institutional design	Very initial work testing two new forms of exchange with demand-side bidding	

the institution. This important conclusion can be summarized as "institutions matter because rules matter and rules matter because incentives matter."[44]

Experimental economics can be classified into at least seven types of usage. See Smith (1994). In this section, the use of these possibilities is linked to the experiments performed to give an overview of the experimental study at a level of economic study.

Table 1.2 shows what aspect of experimental economics is used in which experiment and for what purpose.

[44] Quote from Vernon Smith.

References

Arrow, K. J. (1963). Uncertainty and the economics of medical care. *American Economic Review, 53*, 941–973.

Arrow, K. J. (1965). *Aspects of the theory of risk-bearing*. Finland: Yrjö Jahnssonin Säätiö.

Arrow, J. K. (1962). Economic Welfare and the Allocation of Resources for Invention. *In NBER The Rate and Direction of Inventive Activity: Economic and Social Factors. Princeton Univ. Press (for the NBER).*

Barzel, Y. (1986). Knight's "Moral Hazard" theory of organization. *Economic Inquiry, 25*, 119–120.

Baumol, W. J. (2002). *The free-market innovation machine: Analyzing the growth miracle of capitalism*. Princeton NJ: Princeton University Press.

Coase, R. H. (1937). The nature of the firm. *Economica, 4*, 386–405.

Coase, R. H. (1946). *The marginal cost controversy*. London: London School of Economics and Political Science.

Coase, R. H. (1960). The problem of social cost. *Journal of Law and Economics, 3*, 1–44.

Crouch, D. D. (2008). The rising size and complexity of the patent document research paper. University of Missouri School of Law.

Friedman, M. (1976). *Price theory*, Aldine de Gruyter.

Gambardella, A., Giuri, P., and Mariani, M. (2005). The value of European patents evidence from a survey of European Inventors. Final Report of the PATVAL EU project, Contract HPV2-CT-2001-00013, January 2005.

Hanson, N. (1971). *Observation and explanation: A guide to philosophy of science*. New York: Harper & Row.

Hanson, N., & Humphreys, W. (1970). *Perception and discovery: An introduction to scientific inquiry*. San Fransico: Freeman, Cooper.

Hart, O. D. (1988). Incomplete contracts and the theory of the firm. *Journal of Law, Economics, & Organization, 4*, 119–139.

Heckscher, E. (1919). The Effect of Foreign Trade on the Distribution of Income. *Ekonomisk Tidskrift, 21*(2), 1–32.

Hurwicz, L. (1960). *Optimality and information efficiency in resource allocation processes*. Stanford: Mathematical Methods in the Social Sciences.

Hurwicz, L. (1973). The design of mechanisms for resource allocation. *The American Economic Review, 63*, 1–30.

Kahneman, D., & Tversky, A. (1979). Prospect theory: An analysis of decision under risk. *Econometrica, 47*, 263–292.

Kamien, M. I., & Tauman, Y. (1986). Fees versus royalties and the private value of a patent. *The Quarterly Journal of Economics, 101*, 471–492.

Kaplan, A. (1964). *The conduct of inquiry*. Chandler: San Francisco, CA.

Katz, M. L., & Shapiro, C. (1984). *How to license a patent*. Princeton, NJ: Woodrow Wilson School, Princeton University.

Kaufer, E. (1989). *The economics of the patent system*. Chur: Harwood Academic Publisher.

Kharif, O., & Miller, H. (2011). Ericsson said to weigh entering contest for nortel patents. *Bloomberg Markets*.

Klee, M. M., & Fairfield, C. T. (1998). Where did the US patent laws come from?

Knight, F. H. (1921). *Risk, uncertainty and profit*. New York: AM Kelley.

Krugman, P. (1985). *A 'technology Gap' model of international trade*. London: Macmillan.

Krugman, P. (1990). Endogenous innovation, international trade and growth. In *Rethinking international trade*. The MIT Press.

Lamoreaux, N. R., & Sokoloff, K. L. (1999). *Inventive activity and the market for technology in the United States, 1840–1920*. Cambridge, MA: NBER.

Lamoreaux, N. R., & Sokoloff, K. L. (2001). Market trade in patents and the rise of a class of specialized inventors in the 19th-century United States. *The American Economic Review, 91*, 39–44.

Leroy, S. F., & Singell, L. D., Jr. (1987). Knight on risk and uncertainty. *The Journal of Political Economy, 95*, 394–406.

Nordhaus, W. (1969). *Invention*. Growth and Welfare: MIT Press.

Nordhaus, W. D. (1972). The optimum life of a patent: Reply. *The American Economic Review, 62*, 428–431.

Ohlin, B. (1924). The Trade Theory. *Flam, Harry and Flanders, M June (Eds.), Heckscher-Ohlin trade theory*, pp. 82–213.

Plant, A. (1934). The economic theory concerning patents for inventions. *Economica, 1*, 30–51.

Popper, K. R. (1959). *The logic of scientific discovery*. London: Hutchinson.

Reiter, S. (1977). Information and performance in the (New) welfare economics. *The American Economic Review, 67*, 226–234.

Ricardo, D. (1891). *Principles of political economy and taxation*, G. Bell and sons.

Schumpeter, J. A. (1934). *The fundamental phenomenon of economic development* (pp. 57–94). The Theory of Economic Development: Schumpeter.

Schumpeter, J. A. (1942). *Capitalism, Socialism and Democracy*, New York: Harper.

Smith, V. L. (1962). An experimental study of competitive market behavior. *The Journal of Political Economy, 70*, 322–323.

Smith, V. L. (1968). Optimal insurance coverage. *The Journal of Political Economy, 76*, 68–77.

Smith, V. L. (1981). An empirical study of decentralized institutions of monopoly restraint. In J. Quirk & G. Horwich (Eds.), *Essays in contemporary fields of economics in honor of E.T. Weiler*. West Lafayette: Purdue University.

Smith, V. L. (1982). Microeconomic systems as an experimental science. *The American Economic Review, 72*, 923–955.

Smith, V. L. (1994). Economics in the laboratory. *The Journal of Economic Perspectives, 8*, 113–131.

Ullberg, E., Rodriguez, E., & Stormby, N. (2002). Risk management – from portfolio strategy to value creating system strategy. *Geneva Papers on Risk and Insurance – Issues and Practice, 27*, 467–476.

Vickrey, W. (1962). *Auctions and bidding games. Recent advances in game theory, 1962*. Princeton, NJ: Princeton University Press.

Chapter 2
A Dynamic Microeconomic System Design for Markets in Patents

2.1 Introduction

The goal of this chapter is to develop an experimental and dynamic economic system design useful to study organized markets in patents.[1] The problem studied is the "last step" in a historic transition from personal to impersonal exchange in ideas and some prerequisites for transparent prices on patents, dynamic gains from trade in the rights themselves, and coordination of technology search through prices. The main independent variables in this study are the therefore mechanism designs and patent system strength. An experimental approach is used since the mathematical challenges are currently prohibitive to model such a complex and dynamic system as the patent system and there is no such market to produce data to be studied at this stage although many attempts are ongoing.

The standard example of the use of the patent system is that in which a firm use the system to protect an invention (a technical idea) and exercise a temporary monopoly on the *product*. The economic analysis of such a use has normally proceeded along the lines of divergence between private and social product in an integrated firm hierarchy. In this treatment, economists have largely used the excluding right of the patent and marginal process (cost-reducing) inventions. This is the treatment by Arrow (1962, p. 51) on the allocation of resources for invention, Krugman (1990) in the technology factor of trade, and the approach of Schumpeter (1942) on

[1] This chapter, primarily concerned with the dynamic outcomes of trading patent rights in an experimental economic system, arose out of the study of management of risk and uncertainty through new market mechanism designs at ICES-GMU. See also: Ullberg (2009). Special thanks are expressed to the Savings Banks Research Foundation and Dir. E. Rodriguez for funding this research, Royal Institute of Technology in Stockholm, and numerous personal communications with Prof. V. Smith, S. Rassenti, D. Porter, O. Hart, as well as input on market design from Goldman Sachs. The structure of the chapter was inspired by Coase (1960).

E. Ullberg, *Trade in Ideas: Performance and Behavioral Properties of Markets in Patents*, Innovation, Technology, and Knowledge Management 13,
DOI 10.1007/978-1-4614-1272-4_2,

integrated markets. However, the patent system generally gives two rights, the right to exclude and to transfer or conclude licensing contracts.[2] To this end, the optimal type of licensing contract has been analyzed as a transfer of the right under monopoly conditions in a product market. This is the treatment by Kamien (1992) and others. These treatments are interesting but do not take into account the competitive nature for technology. The second right establishes a *competitive* market in technology (not simply a seller's monopoly), with demand-side bidders (competing for the technology). The conclusion of both these approaches has been the static gains from a patent system in terms of incentives to allocate resources for invention of cost-reducing inventions and contracting incentives in terms of private and social gains, leaving the dynamic gains from this often highly competitive technology market to further research.

It is my contention that discussing the patent system primarily in terms of a *static* product monopoly for integrated firms or markets may miss essential trade gains coming from trading the rights themselves between specialized actors in a dynamic and *competitive* market, with transparent prices. The static studies of the patent system may even lead to results which are not necessarily consistent with respect to how these rights are used and their effect on the structure and performance of the economic system (the model changes when competition is introduced).

A experimental and dynamic economic system is developed, expanding on the static microeconomic model for experiments of Smith (1982),[3] to study three initial areas of interest: (a) competitive demand-side bidding under different mechanism designs for public price formation, (b) dynamic gains in the use of technology among specialized agents enabled by the patent system, and (c) price signaling between demand for technology and search for economically useful technology. Different auction mechanism designs and presumed legal validity of patents are the main independent variable and the investigation of the performance and behavioral properties of these mechanisms is the focus of the design.

An informal price theory is outlined for a linear contract on patents and initial hypotheses stated. These questions begin to study the dynamic aspects of this exchange of which there are many more possibilities which I leave for future studies.

I will begin with a summary of the static approach and then give an overview of the change in economic structure that has taken place since the beginning of the patent system in 1474. Followed by this historic and theoretical re-framing of the patent system as an exchange system is a detailed description of the dynamic

---

[2] WTO TRIPSs agreement, Article 28:2, "Patent owners shall also have the right to assign, or transfer by succession, the patent and to conclude licensing contracts."

[3] The approach is based on Smith with institutions, I, and an economic environment, $\mathbf{e}$, with changeable tastes for technology through a discovery/sampling process with some limits: $\mathbf{s}=(\mathbf{e}, I)$. Here a legal environment, P, is added and the ability to "split" contracts according to claims on patents for limited exclusive use in different product markets. The contracts are tradable in a primary and secondary market.

economic system design, and its components, intended to be useful for experimental studies. This part includes a section on institutional design for mechanism designs differing in the integration of both information and rules, with hopefully quite different incentives to express willingness to pay/willingness to accept, forming the basis of the proposed initial experimental studies. An outline of a price theory then follows. Finally, initial hypotheses to begin the investigation of the exchange system design are formulated.

2.2 Institutional Learning in History and the Two-Dimensional Nature of the Patent System

2.2.1 A Competitive Market in Technology

The traditional analysis tends to obscure the nature of the decisions that have to be made when trading patents in their own rights. The question is thought of as one in which a firm is investing in risky research and development activities, "production of information[4]," and should decide whether or not to patent the inventions in order to use them while protecting the ideas from being copied (defensive use) or to "block" competition from entering a business "too close" to its own business (offensive use); this is a problem of an anticompetitive nature. But this viewpoint is wrong from a trade context view. Here, we are really dealing with a problem of *competitive* market nature. This market can also be thought of as a technology market for producers (Innovators) and investors (Traders) deciding to invest in an already patented technology. This *producer* market is developing in a direction that can be described as moving from *personal* exchange through negotiations, e.g., licensing, cross-licensing, to *impersonal* exchange with prices through competitive bidding in institutions with rules based on more standardized contracts.

However, pricing a patent is not the same as pricing potatoes or shoes. The patent has *two* sources of trade value based on its possible dual strategic use: investing in the useful technology or blocking someone else from doing so, making the competitive trading problem a two-dimensional one.[5]

---

[4] Prof. David E. Andersson, pointed out the difference between information (codified knowledge) and tacit knowledge (skills, capabilities). This is dealt with in the patent application in the "disclosure" section where the invention must be described so that "a person skilled in the art" can reduce it to practice.

[5] In the optimal patent licensing literature, one-dimensional markets are investigated, often studying either a fix fee licensing price or royalty price of a linear contract, negotiated in an auction, not both simultaneously.

Blocking, or "sitting on," a contract is very different in terms of the risk[6] involved compared to investing in innovations and using the technology for products. The blocking value is also a value that is more "common" among agents. It does not matter who is "sitting on" it, it is still excluding others from performing certain competitive actions. Blocking allows the holder who is blocking to continue to earn money from producing based on its current portfolio of patented technology. This value is there even if the producer does not hold the contract, but remains strategically exposed to further development in the area of the protected technology, a value that can increase or decrease with time and be an investment even for a Trader.[7] Up to 50% of patents who are renewed may be used in this way.[8] ("Sitting on" is therefore similar to trade secrets, but tradable, however much less certain.) The one that can extract the value however is one that can sell such a contract to a producer who can realize its blocking value. Sitting on such contracts and then selling them is therefore also a strategy possible for financial traders not only Inventors. The blocking use by a producer allows depreciation of current assets and/or timing of new innovations given the *market access* the firms currently have. The timing horizon, given the competitive environment and current assets, is thus related to the cost of capital of a firm in deciding between projects [using discounted net present value (NPV) of cash flows from operations]. The blocking value can therefore be seen as protecting the market access of current and future products from more competition, extracting more money by limiting competition from new product innovations or product "cannibalization."[9] There is a similar strategic use in "corona patents."[10]

---

[6] The word "risk" is here used for risk and uncertainty. Compare Knight (1921) for famous distinction between risk and uncertainty. The use here is that an insurance market can be created for risks – risks can be given a number and are "insurable" – and uncertainty that has to be *managed* (through people, systems, and intellectual property rights). The view on uncertainty appears to differ from Knight in that management and profits are not a complete lottery. At the INSEAD business school, Prof. in statistics S. Makridakis held for true that flipping a coin and hiring a manager to make the decisions were statistically equivalent in outcome, thus possibly supporting Knights view. Compare also prospect theory with "insurance" and "gambling" as decisions under uncertainty. Management of risk and perhaps in particular uncertainty appears to be an area of increasing importance to study.

[7] Here, we are concerned with investors who take positions in technology as a business where the business model is to maximize the use of the technology. These companies sometimes are investment companies with patent pools or portfolios, which may give market access to the owners at more reasonable prices. This practice is not new and has been practiced strategically by competing producing companies, but it is fairly new when it comes to a separate business. These are not so-called "patent trolls" whose intention is to extort rather than create value. The problem with "trolls" is related to creating incentives for innovations.

[8] *Source*: See Licensing Executive Society's statistical surveys (Razgaitis 2006, 2007).

[9] Cannibalization is a concept in management where own new products take market share from existing products, "cannibalizing" on the market share of the existing products.

[10] Corona patents refer to patent that are granted for invention building on, or around a basic patented technology. Once the basic technology is free, the corona patent holders exclude basically any future development for the original patent holder. The most famous case here is Xerox versus the Japanese photocopy industry which was able to develop new applied technology and once the basic patents were expired, caused a severe blow to Xerox.

Innovating new products by investing in using the patents is much more a private value to a specific company or industry. Here, the technology is being used to *create* a new competitive advantage through product innovation together with current technology portfolio of the company. The risks with respect to market access are here higher, since developing a new product to create more market access is typically more uncertain than producing current product mix.[11] However, since competition and needs change, innovation is a strategy companies need to assure long-term market access.

2.2.2 Allocation of Risk Bearing and Risk Sharing

The intent of the patent system is to create incentives to innovate not only to invent, i.e., create *useful* inventions.[12] If trade in patents is to be included in the economic equation, its purpose should be to achieve a more optimal allocation of the inventions with respect to managing the market access risks in the innovations. If measuring market access risk by volatility in revenues (sales), the volatility in current sales is typically much less than from new products, making the risk of blocking lower than investing. This suggests that blocking is a less risky strategy than investing. In trading this right competitively both the investment value *and* the blocking value of patent claims have therefore to be taken into account. The contention is therefore that the instrument traded has to be two-dimensional, allowing the expression of the two risks independently. Such an instrument is the *linear contract.*[13] I will come back to the reasons for using the linear contract in Sect. 2.3. For now the conclusion is that to encourage organized trade of patents, its two-dimensional value would need a two-dimensional contract to allow efficient trades with respect to its economic use, and to manage market access risk in customer relations and the economic system as a whole (i.e., incentive compatible outcomes may best be achieved through two-dimensional, "multitariff," markets).

2.2.3 Historic Development

The development of organized trade can be historically framed by trade in technology that shifts from *personal* exchange to *impersonal* exchange.[14] See Fig. 2.1.

---

[11] Product mix is a concept in management where the market access risks are managed by not having "all eggs in the same basket." Different products can target different market segments to spread the risks in more than one product market. A good product mix strengthens the relations with customers and reduces the market access risk.

[12] Plant asked the question that the crises in the 1930s might have a relationship by creating "too much invention of the wrong kind" Plant (1934, p. 51).

[13] For the linear contract literature, see Tirole, Lafont, Kim and others.

[14] For a broad discussion of the theme of "personal to impersonal exchange," see Smith (2003, 2007). Also Grief (2002).

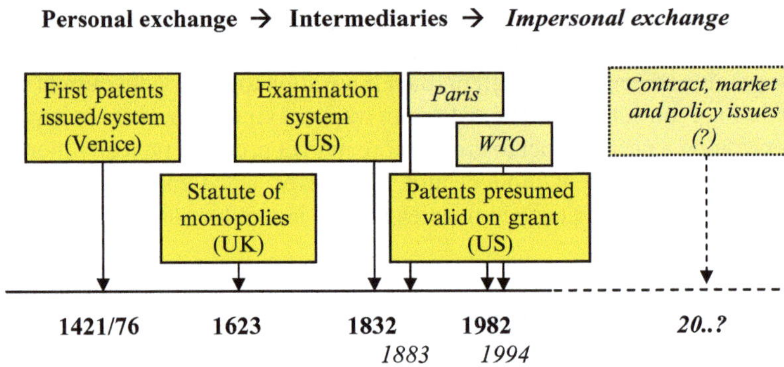

Fig. 2.1 From personal exchange toward impersonal exchange in patents

Personal exchange refers here to the current trade in patents, mostly licensing.[15] This use has seen a rapid growth in the last 20–25 years, especially in the USA,[16] passing one trillion dollar according to widely accepted but rough estimates. The exchange may include intermediaries (patent lawyers, accounting firms, investment and commercial banks, etc.) in the contractual and valuation side of these transactions. Intermediation is not new and has been studied by economic historians with focus on transactions, notably the effects of the US patent reform in 1836. The reform added the "invention" of the examination, in addition to simple registration which had previously left validity of uncertain claims up to the courts. Now patents could be *presumed* valid upon grant which led to the evolution of a new industry of specialized patent *Traders* emerged (Lamoreaux and Sokoloff 1999; Lamoreaux and Sokoloff 2001). In the 1400s, the first patent system was established in law in Venice in 1474 creating "industrial monopolies" making trade secrets public. In 1623, in the UK (England 1623), these "industrial" monopolies granted by the Crown were abolished and replaced by 14-year patents on products and processes

[15] Today, current exchange in patents takes place between companies for a fee or cross-licensing or both, within companies, for policy reasons, tax reasons, cost reasons, within pools of patents, to simplify, reduce patent "stacking," and cost of market access, within standards, typically under RAND conditions, patent portfolio securitization, a bond that in the rare cases so far yields licensing revenues, other forms of bilaterally negotiation and contracting, or initial bankruptcy or specialized "patent auctions" in one dimension (fixed fee). RAND (reasonably and nondiscriminatory), terms is an agreements often used with standards to allow inclusion of patented technology in a standard, making it more attractive and useful, without excluding any market actor. This is a common activity in the telecom industry where interoperability is a key. The competition then moves to the product side with design, trademarks thus other IPRs and at the same time the patent owner can secure a "reasonable" royalty from the technology. This is such an important field that Sony chose to make the Sony Ericsson joint venture on the handsets since it was too costly to license the technology (*Source*: a public interview with the CEO of Sony at the time).

[16] In the USA, a federal circuit patent appeals court was introduced in 1982 as well as a rule change comprising that patents granted by the patent office should be considered valid when challenged, creating a more asset like right of patents.

which, as it appears, created the two-dimensional nature of the contract by opening the door to blocking since the value of the previous "industrial" monopolies were in the production. In 1982, the presumption of validity, previously given to the society, was returned to the Inventor, who is taking the risks in the inventive process, effectively creating a situation comparable with ownership of physical assets in terms of presumed validity.[17] In the international arena, the Paris convention from 1883 turned the patent system into an international system (through a priority year and national treatment) and the WTO/TRIPS agreement declared "minimum standards" for all member countries (a contested agreement) (WIPO 1883; WTO 1994). A theme that appears to be motivational for this institutional evolution is the importance of economic development through technology *trade* (import/export from the city state level to the national level to the international level).

Trade secrets have thus received public protection, public disclosures, and become tradable in an *impersonal* sense, over a process of more than 500 years. What appears to be the current direction is *public* prices: making technology trade an even more integral part of efficient economic development. The focus for the economic system designer is therefore impersonal exchange imbedded in a testable market institution to study the prices and gains from such IP trade. The buyers and sellers will interact through rules set by the "intermediary" (the institution). As a result, bidding and clearing prices will become *public*. Such trade with *public* prices accentuates the *dynamic* side of an economic system where many Inventors can sell, with help of Traders, to many Innovators for multiple technological usages.

The question that has to be decided is therefore how the competitive prices of a two-dimensional contract on patents ought to be determined (theoretical price) and under which institutional arrangements we can observe the most accurate prices (with respect to predictions) as well as possible gains from trade given the presumed validity (legal institutions).

2.3 The Dynamic Microeconomic System Design

The impersonal exchange testable market studied is part of a microeconomic system design experiment using trained cash-incentivized subjects.[18] The economic environment, e, the institutions, I, and the legal environment, L, are designed to

[17] In the USA, the presumed validity went from 38 to 93% during the 10–20 years following the 1982 decision.

[18] One here commonly makes the distinction between "naturally occurring markets" that can be observed in the economy and "design markets" which are constructs whose properties are examined in a laboratory experiment or in a game theoretical mechanism design analysis. The contention that the designed system is a real microeconomic system allows for parallelism between the designed and the observed economy at some level of abstraction, and is supported by the fact that real people make real economic decisions receiving real cash payments for their gains. If the payments are "salient" enough then other preferences are argued to be suppressed and the profit incentive dominate behavior, thus cash-incentivized subjects. See Smith (1982).

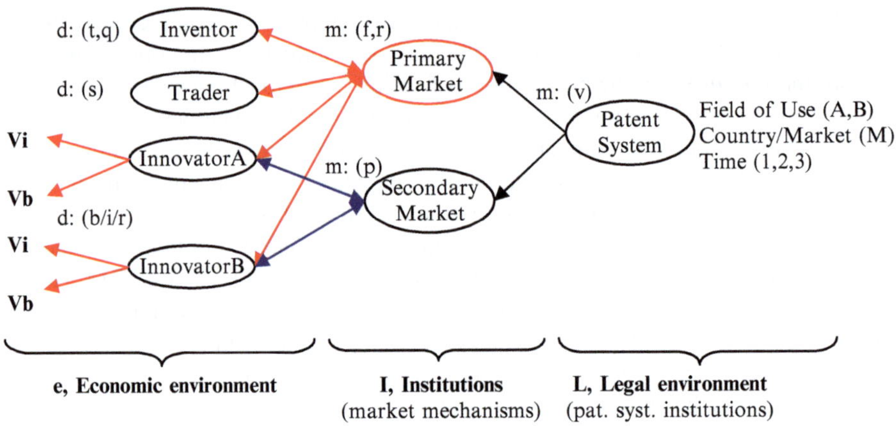

Fig. 2.2 The dynamic system design: *s*, to study prices and gains of the patent system during patent time. $s=(e, I)$. $I=$I1, I2, L. Vi/Vb = induced value ranges for investment/blocking options. $m(f, r)$, $m(p)$ = ask/bid. $m(v)$ = validity message. $d(t, q)$ = tech.focus and quality decision by Inventor. $d(s)$ = decision by a Trader to split contract in multiple usages. $d(b/i/r)$ = Block/Invest/re-sell decision by Innovator

capture the *principles* of the patent system, including the dynamic effects: the two-dimensional nature of the pricing problem, the decisions that agents specialized by type of activity, invention, trade, and innovation can *do* dynamically under induced uncertain private values of the patented technology, and only presumed validity.[19] Figure 2.2 gives an overview of the designed system.

The Inventor creates a contract to trade having a certain technology focus (t) and quality, q, using decision d:(t, q). Traders can split, s, using decision d:(s).The agents send messages, m:(f, r), expressing the fixed fee (f) and royalty (r) in the primary market. In the secondary market a single price, p, is negotiated using message, m:(p). The patent system has a certain presumed validity, v, which is sent through message m:(v) to the agents. Innovators can block, b, invest, i, and receive a value Vb or Vi, using decision d:(b/i). In periods 2 and 3, Innovators can also re-sell, r, using decision d:$(b/i/r)$.

By studying the system at a level of institutional principles, attempting to choose the environmental design parameters and the uncertain private values to approximate what exists today in the economic and legal environments, we can examine in a repetitive and controlled manner, prices and adjustment processes, and the dynamic

[19] All systems are based on principles. If the principles are working, we have a functioning system. We can study how these principles are working in this complex system of trade in IP through laboratory experiments and attempt to draw inferences from the outcomes with respect to these principles. Uncertainty in private values is introduced by given the values in a range from a random distribution rather than fix values. Under rational expectations, the value "used" by a subject making decisions on prices would then be the mid-point/average.

effects of the designed institutions on the system as a whole (Pareto gains). Using this approach, policy and strategy conclusions can hopefully be more easily drawn.[20]

Since the patent system is used in a complex way, I propose to start the description of the system by a description of the parameters and assumptions (legal environment). The decision-making inherent to the principles is then discussed (economic environment). Finally, a simplified mechanism design is presented (institutional environment).

2.3.1 The Legal Environment

To introduce the patent system as an independent variable, the typical *economic* environment, including economic agents, commodities (and resources), and certain characteristics of each agent, is extended to include a *legal* environment. This makes it possible to vary the presumption of validity of a patent which affects the risk in trading patents. In the environments tested, the presumed validity is varied between "high" and "low." The high value used, 93% probability of validity during the life of a patent, is chosen to match current approximate levels of US PTO patents.[21] The low value used, 38%, is chosen to match the level of US patents prior to the 1982 change in burden of proof (a patent granted by the patent office is now presumed valid when challenged in court, giving the benefit of the doubt to the Inventor) *and* to match current approximate level of EU (EPC) patents[22] (currently also approximately 38%)

---

[20] This total system approach has been advocated by Coase (looking at a problem "in total and at the margin," (Coase 1960) and others as a way to draw conclusions for policy that would have some relevance to the actual situation.

[21] *Source*: EPO's estimates of validity for USA (93%) and EU (38%) around year 2000.

[22] Note the important distinction between the PCT route of a central *filing* system for applications and the national (e.g., USA) and regional patent laws (e.g., EPC) granting the rights. The "international" patent system is still the Paris convention (1883) granting applicants national treatment and a priority year to file in each and every nation of the treaty, i.e., a system based on these two *principles*, and not a 'harmonized' patent law. The UN's PCT system, administered by WIPO, has for the last four decades been developing toward a real international system. Recently, through Chap. II of the PCT, the system approaches the status of a patent *granting* system by issuing a preliminary reports on patentability, moving the PCT closer to an international patent system. The main "obstacle" here seems to be a difference of principle between the US system of "first to invent" plus a "grace year" to file a patent application for the invention and the EPC, JPO, and rest of the world systems (ROW) of "first to file." The US system gives the inventor the patent in his own rights (the inventive activity). The ROW system is an system where the state decides the priority date by receiving an official application (the administrative activity). The US system thus gives the advantage of being less administrative by principle, however, at the price of an elaborate case law to prove the date of invention, if the patent is challenged, creating uncertainty of validity. Most recently (2009), a first-inventor-to-file system is proposed in USA to "bridge" some of the differences and making the US system also a filing system. The current trend toward harmonization of patent laws challenges the historic policy use of the patent system as creating *national* competitive advantage. Whether "competing systems" or "harmonized patent laws" is the best for world trade and global economic development is here a question. Here, the focus is on the trade once the patents are granted and any uncertainty in legal validity is parameterized by "presumption of validity" as a percentage that is uniformly distributed, $U[0, 100]$.

that have the "pre-1982" US burden of proof doctrine (a patent granted by the patent office is presumed not valid when challenged in court, giving the benefit of the doubt to the society). This shift in the benefit of the doubt to the agents who are actually taking the risk in purchasing technology may have been the necessary step to make the patents truly tradable.[23] The system thus provides the possibility to study prices, usage (innovation/blocking), social gains (and possibly other economic consequences) for the alternative legal patent regimes (a question of principle).

The innovative use of the patent right (technology) can be increased through derivates made possible through the patent system's design: fields-of-use (claims), geography (designated countries), and time (excludable years). "Splitting" the right into several fields-of-use (for different product innovations) is configurable in the experimental environment. The periodic renewal of patent rights fee used is set to zero in the initial studies but remains configurable as an institutional parameter for future study.[24] The economic time horizon is set to three periods, representing the typical 20 years of legal value.

2.3.2 *The Economic Environment*

2.3.2.1 The Economic Agents

The economic environment is a fixed environment in this experiment; the independent variables are the institutions and the legal environment. To study the transfer of rights aspect, we can distinguish between three activities: inventing (creating and selling the rights), trading (buying and selling, to maximize investments in the rights in potentially multiple innovations), and innovating (buying and investing in using the rights). These activities are all more or less present in today's *personal* exchange but the activities are still *mostly* carried out within a single firm (coordinated in a hierarchy). The rapid increase in licensing of patents and other IP during the past two decades is changing this. In the experimental environment, these activities are executed by the specialized economic agents, each performing only one of the tasks,

[23] This shift is often referred to as the beginning of the "pro patent" era with emphasis on the claims (the granted rights). Previously, there was a "pro society" era, with an emphasis on the disclosure of the invention (what is made public about the invention).

[24] One way of dealing with renewal fees are exponentially increasing fees over time. This is in practice at, for example, the EU with EPO. The economic argument is that to create incentives to innovate, the cost of renewing the patents should be taken out on the innovations when and if they are likely to be profitable for the inventors. This takes place years after patenting the invention and the innovation is introduced in the market. Recently, there was a proposal in the USA to introduce increasing fees over time. Protection for inventions that are economically valuable are then renewed by its owners and others are dropped, potentially making access to technology easier and incentive to develop valuable technology increases. How much Innovators should give back to society in monetary terms is a contentious issue but the issue here is also about creating economic incentives to use the inventions in innovations. This second point deserves a more than a footnote though.

making it possible to study different pricing mechanisms for patents (coordination in a market with prices that are public). The specialized economic agents will be referred to as: Inventor, Trader, and Innovator. This organization of agents is motivated by differences with respect to the risk that has to be managed (uncertainty in market access) in each type of activity, thus they may have different costs of capital (referring to CAPM or the intertemporal ICAPM with "betas" for validity, etc.).[25] This assumption can be seen as supported by the fact that venture capital typically asks around 30% in return, financial institutions 3–5% (3.5% for long-term return adjusted for inflation and taxes[26]), and shareholders in listed companies 10–15%, dependent on industry. In the experiment, 30–5–10% are used for the respective categories of agents. This key economic characteristic of the design allows a new type of agent to be introduced who invests in patent rights, the Trader. The specialized Inventor firm is also a new kind of company, operating under a high-risk business model, possibly requiring more capital to succeed through trial-and-error R&D projects. This is an increasingly important business model for many inventive start-up companies today. The cost of capital can then be calculated for different systems, indicating changes in the systemic risk.[27] A strategic problem with the specialization may be that not all information with respect to the patented technology can be transferred, like, for example, know-how.[28] One way to deal with this could be to also add a second contract option that guarantees transfer know-how to the buyer of the contract traded. Future inventions in the same field may have to be bundled in order to not to Block a potential buyer in the future. Institutions which facilitate bundling are possible but seriously more complicated and will be left to future study.

[25] This assumption is in fact a tentative *proposition* that the boundary of a firm may be organized in terms of managing genuine *uncertainty* specifically related to ideas (market access). This means that there is a proposed difference between managing uncertain market access related to *new* sales (revenues from new innovations) and cost (revenues from old innovations). This may have similarities to "cost of carrying out a transaction" or "marketing costs" (Coase 1937), or differ in terms of investment costs, expressed by the cost of capital, as a determining factor for firm organization, reflecting risk in their long-term market access of a firm and how well they manage these risks through people and systems (in particular, information systems). The argument is here that the proposed division of activities actually is based on difference in the uncertainty of carrying out that activity. This would be reflected in a competitive market priced access to capital, the cost of capital. The experimental economic environment includes this parameter as a controlled variable by the experimenter, but experiments making it an independent variable could be designed.

[26] Refer to personal communication with B. Lev.

[27] The systemic risk here refers to the capital asset pricing model (CAPM) and intertemporal CAPM (ICAPM) method to calculate the *systemic risk* (the alpha) of a market and a *project risk* for a project (the beta(s)). This is a mere calculation exercise in the experiment, based on endowed typical cost of capital of the agent types, given the actual allocation of risk bearing among subjects of different agent types. As mentioned elsewhere, the design may be used to make this a more endogenous decision in future experiments.

[28] Information can and is also kept private as trade secrets, exposing the buyer to further inventions by the sellers. Such behavior of course drives down the price for an invention. This problem is thus related to the disclosure requirements of the patent, an issue of policy.

The primary purpose at hand is to limit the study to possible gains from the trade of *publicly* disclosed information through the patent system, which may exclude some important second-order strategies from the study that may be of consequence in naturally occurring markets.

Inventing typically includes extensive research often performed in an organized fashion by whole teams of highly educated professionals (up to 40% of Inventors have a PhD degree according to a recent EU study[29]) working with colleagues internationally or over the Internet which has turned research into a global activity today,[30] and patenting to receive the patent right for granted claims, strategically selecting validation states in the process.[31] The "prepatent" phase involves activities and decisions of basic research by universities and governments, joint-research ventures, "patent races," etc., where the patenting policy may differ between actors, thus not all commercially useful inventions are patented, for a number of reasons (military use, patentability, detectability, also other IP may be favored, like trade secrets, etc.). The patenting phase can be thought of as a process producing "patent products" for certain country markets that can be used for innovation purposes. This practice creates a "portfolio" of patents which represent "technology assets" of a company. The *patenting* process is highly specialized and external "patent lawyers" here do much of the work today thus then "inventor agent" is at the minimum includes the Inventor and a patent lawyer.[32] This service is a rather costly and comes at the initial stages and is included in the model as a *cost* of creating a contract (not a special agent). A simple portfolio of one (1) contract is included in the design and only one market (validation state). Neither the "prepatent" nor the "postpatent" use of technology is included in this design. The contract cost varies depending on the "quality" of the patent. High-quality patents cost twice as much as low-quality ("standard") patents.

[29] Refer to European report on invention, Gambardella et al. (2005).

[30] This "large corporation" situation does in no way exclude the individual inventor or SME, who have similar processes, in particular global due to Internet. Also the majority of patents are granted to large corporations who typically have research organizations. It takes research to be in the patent business. All research frontiers are in practice globally accessible (within their publication sphere: private, public, government) very shortly after discovery, making research a global activity, which is the main point here.

[31] Validation strategy is very important from both a competitive and a cost perspective. Validation is typically decided by three criteria: presence of customers/competitors, presence of possibly competing manufacturing facilities, presence of competitive research firms (since the patent system also includes excluding rights for research). This very complex *market-oriented* approach technically results in a number of validation countries for a patent. Since the cost of a patent is about one million dollar during a patent's life (*Source*: Private US law firm), and even the largest companies have limited patent budgets, this strategic validation typically results in about 7–8 countries or even less, across all technical areas, the exception being pharmaceuticals who see a competitive threat in "every local pharmacy," thus having much wider country coverage (*Source*: EPO communication). Previously for many companies, the absolute *number* of patents was an important strategy. This is changing and a renewed focus on "quality" is made. Those patents are also more tradable which may be the key a motivating factor.

[32] Prof. Rassenti suggested that this agent was very important.

Innovation typically includes extensive market research, product development, marketing, sales, product life cycle management, product mix strategy, and patent portfolio management as well as the use of know-how and technology that can be patented is expired or nonpatented. The innovation phase is the "using phase" of the patent product. "Spill-over" effects and other "post- or prepatent" effects are not explicitly included, except the "signaling" of the value of a contract that takes place through the market prices.

Trading in patent products today typically includes different kinds of licensing agreements: the purpose being to free up present (licensing) and future (potential licensing) market access, and for specialized actors, like patent lawyers, enhance the possibility to "split" the patent products into multiple fields of use (claims), geographical markets (validation), time, exclusive, or nonexclusive for multiple licensing of the same technology. Traders have incentives to "maximize" the available use of the patents across all Innovators, making them of key importance to improving gains from trade. In the experimental model, the dynamic adjustment comes from Traders "splitting" one of the contracts into two "fields-of-use." A possible future extension of the design includes mechanisms for endogenous "bundling" of patents (in addition to the splitting by Traders). This could be realized through combinatorial mechanisms and rules but requires an adoption to a two-dimensional linear contract.[33]

2.3.2.2 The Security Traded

In Sect. 2, it was argued that the contract traded had to be two-dimensional to accommodate for the simultaneous investment and blocking values of the patent right. The experimental environment will use a linear contract with two components: a fixed fee plus a royalty contingent on sales (revenues).[34]

The left part of Fig. 2.3 illustrates that the linear prices are made up of combinations of fixed and royalty payments. The figure to the right illustrates the payment from a

[33] Design combinatorial auctions for one-dimensional pricing have been tried in the laboratory and applied tests by Porter et al. (2003, 2009).

[34] The contract to use has been the subject of considerable attention and challenge. Prof. V. Smith, first drew my attention to the fact that a "one-dimensional" contract is not different to that of trading potatoes or shoes. This lead the thought to question the contract itself, whether it would be suitable to trade IPRs. A key input was given by Prof. D. Porter who saw similarities between the risk-sharing problem discussed and "sharecropping." That literature search led to Stiglitz (1974), incentives and risk sharing in sharecropping, which in turn led to the literature on linear contracts (Tirole et al.). Also, input from an inventor suggested that a fix+royalty would be preferred although rarely possible, strengthened the author's belief in using a linear "two-dimensional" contract. When the two-dimensional nature of the value became clear, it all seemed to explain the rationale behind why inventor agents prefer such a contract and buyers accept such a contract in some cases. Special data from a unique source on patent licenses confirmed that linear contracts with some fix payment were used in 50% of the cases and only 13% with an annual minimum payment, a truly linear contract (*Source*: RoyaltyStat Inc.). The observed use of contract and the theoretical proposition based on the management of market access risk thus coincide to some extent.

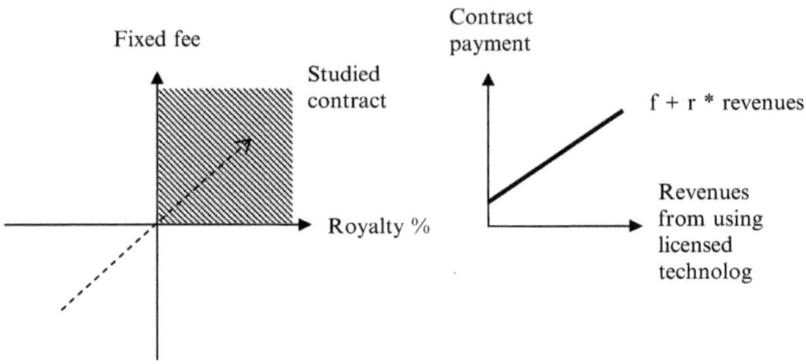

Fig. 2.3 The tradable linear contract: components and payouts

holder to an issuer for using the rights in innovating for blocking. The contract is tradable, meaning that a holder who once bought it (in a primary market) can re-sell it to anyone at will (in a secondary market). The patent itself is not traded, its ownership remains with the Inventor, but the contract is licensing all or parts of the granted patent rights in the patent claims. The contract is binding. This construct makes it possible to "split" the rights into several contracts as described in Sect. 2. It is thus the *marginal* value of the technology that is priced.[35] This leads to a system where the total value, given by gains from trade, and the marginal value can be studied.

A key aspect of the contract is that it expresses both risk *transfer* (the fixed fee) and risk *sharing* (the royalty) between buyer and seller. Thus, in the contract, risks are allocated through the royalty. This may be seen as an alternative to the Arrow-Debreu security (Arrow 1962) of optimal allocation of risk through derivate markets, one for each "state of nature."

Another aspect of the linear contract is that it can theoretically be extended to allow negative values for the fixed and royalty components, which would then represent common usage of financial capital similar to that of debt financing of firms. For the purpose of this experimental environment, only positive values are explored.[36]

[35] Recently, the federal patent court in USA changed the principle on which payments in infringements cases should be based from the "total" product value approach to the value that is attributable to the product from the infringed technology. This could be a recognition of the fact that 100 ds and sometimes 1,000 nds of patents are used in a single product today and that they contribute individually. This could be seen as a "marginal value" of technology to a product.

[36] A positive fix and a negative royalty are similar if not equal to a venture capitalist lending money to as start-up and asking a royalty in return. To avoid moral hazard related to the use of funds, the VC also takes ownership in common stock a board seat, etc. The capital transferred is used for innovation purposes. A negative fix and a positive royalty is similar to a spin-off venture where the mother company is asking for cash back and at the same time paying a "performance fee" to the company for using the IP. Also here common stock are often used to have control over activities. Neither of these may be voluntary due to the moral hazards (additional contract dimension are needed to enforce investment in the contract and not blocking it such as common stock or joint ventures). A negative fix and royalty is similar to a firm financing all costs through internally generated funds and debt, i.e., the hierarchical model.

In extreme cases such contracts can be interpreted as if an Inventor–Innovators (the hierarchical model) are bribing other Innovators to produce less to realize monopoly product pricing (Katz and Shapiro 1985, p. 26). However, we are here concerned with a specialized agent model where both dimensions are negotiated and there is *competition* in technology. With positive component values (as studied), the contract could possibly be seen as creating a new use of capital. A Trader can, by taking a position in a certain contract and split it, finance the time between invention and innovation and gaining from a potentially increased use of the technology. This trade potentially reduces the market access risk for the Trader's position and allows longer-term financing to be made (the 5% mentioned above). This amounts to a new use of capital, to fund *productive* patent portfolios, or to trade in technology.

There are two reasons for the choice of the linear contract in the model: empirical and theoretical. First, the linear contracts are the most commonly used in private exchange (although they are typically nontradable). A fixed fee is typically paid upfront followed by an annual royalty typically related to revenues (sales) or number of units sold (unit sales). Patent renewal fees are typically paid by the buyer (the presumed user of the patent). The rationale for this contract type appears to come from the moral hazard problem with respect to its usage. If the contract is *never* used in a product, the Inventor gets nothing if there was only a royalty. This is a problem commonly expressed by individual Inventors. Their patents end up in the "drawers" of the buying company due to "changing priorities" of the buying firm or no intention to do anything in the first case except "Block" a possible competitor. The fixed fee, it is argued, creates incentives for the buyer to *do* something with the contract (presumably to recover the fixed fee). Without this fee, there may be a little action. However, it also appears that it is getting increasingly difficult for Inventors to successfully negotiate a fix fee at all,[37] thus reducing the incentive effect argued for.

Second, a linear contract covers the problem of two-dimensional *value* having nothing to do with the moral hazard. I propose to illustrate that here in summary, then more fully in Sect. 4. The argument is that a patent's blocking value is formally similar to an insurance contract, in which a *cash payment* is typically exchanged for insurance to cover irreversible losses, *transferring* the risks from the seller to the buyer. This is similar to what the blocking value does for a holder: it "insures" him against irreversible losses of market access (sales). The investment value is different in that a contract used in innovating a new product is accompanied by risky investments in development and marketing. Increasing the market access or keeping long-term market access is not as "predictable" as in the case of blocking to keep existing market access (typically accurately predicted by companies and analysts alike). A risk-sharing agreement over time is needed. Here, the proposal is simply to treat such use as investment in a project and value it at its NPVs minus a royalty on "success" (revenues), minus what is paid in fixed fees, discounted at the project's

[37] Refer to personal communication with Karl Derman, Polhemprize (2007), one of the Sweden's most successful serial inventors, who has jointly with major Swedish companies developed a range of products with global reputation.

cost of capital (a project beta[38]). This reason can be summarized as managing short-term market access by limiting competition by blocking using a cash payment (which is rather certain and can be based on *expected* revenues in an insurance like manner), and long-term market access of increasing competition by investing in new innovations based on the acquired technology (which is rather uncertain and therefore the risk is shared by a percentage of *realized* revenues). Finally, due to the intertemporal nature of the problem, preferences and market conditions may change during the life of the contract and the owner may wish to make changes with respect to technology portfolio. The tradable linear contract can be seen as an excludable, nonrival "security" that is traded under presumed validity of the underlying patent rights. Given these considerations, the linear contract appears to handle the risk mitigation requirements observed in the world.

2.3.2.3 Agent Characteristics

Inventors can (endogenously) decide to (1) enable Traders to be buyers just as Innovators by investing in a quality contract and (2) select a certain technology focus for the contract (used only in the second experimental environment). Traders have the technology (knowledge) to split a contract in two noncompetitive fields of use and can, if they win a contract, sell the two split contracts to two Innovators, one from each noncompetitive "industry," reaping multiple gains. Splitting increases the market access of the technology or reduces excludability in the absolute sense, lowering the risks for the Traders. Innovators who win the contracts in a primary market then decide whether to use the contracts for investing or blocking. The contracts are re-tradable, and an Innovator can re-sell a contract in a secondary market for whatever reason to another Innovator who has the technology to produce products for which the contract is useful. The experimental environment therefore allows the study of how the fixed fee and royalty (f, r) evolve in pricing the linear contract given the investment value and blocking value, legal environment, and institution. The systemic gains created by (1) coordination of investment in a quality contract and a Trader splitting it, and (2) price signals on value of a certain technology focus can also be studied, given the legal environment and institution.

2.3.3 The Institutions

The last components to be defined in the economic system after the economic and legal environments are the primary and secondary market institutions. An important

[38] Project Beta refers to the capital asset pricing model in its temporal (CAPM) and intertemporal (ICAPM) form.

weakness has been that current price theory, patent licensing studies (game theory), and past experiments have been primarily concerned with negotiating one-dimensional contracts under different personal and impersonal market exchange mechanisms like bilateral negotiations and auctions. The theoretical and experimental challenge at hand is to devise and test an institution to trade and price a two-dimensional contract that meets the demands of the risky environment in which Inventors and Innovators are imbedded.

The auction literature for fixed price auctions was pioneered by Vickrey (1962, 1976), Milgrom and Weber (1982), and others. Since risk neutral agents are generally assumed, a special study of risk averse *buyers* was explored by Maskin (1984). Risk behavior of buyers as a function of levels of capital "cushion" in companies was studied by R. Radner. The finding was that when capital is thinning out, a risk-*taking* behavior may replace the otherwise risk adverse behavior observed when capital is sufficient (Radner 1995). This behavioral issue may be of importance to trading risky patents. Work on linear contracts has been done by Tirole and others but the auction itself is performed in one dimension (Laffont and Tirole 1987). Patents ("contracts" with the government) have similarities with incomplete contracts since validity is only presumed prior to a challenge in court. Incomplete contracts, without reference to patents, have been studied by Hart and Moore (1988), Hart (1988), and Tirole (1999). Hart and Moore looked at the renegotiation aspect of such contracts and its relation to the theory of the firm (coordination in a hierarchy).

The literature on "contract bidding" has explored two-dimensional contracts, but only one dimension is typically negotiated here also (Samuelson 1986, 1987). The other is fixed ahead of time (which can be seen as a kind of separate intertemporal negotiation). This is equivalent to the seller (or buyer) setting a reservation value on a contract, a feature introduced in one of the auctions in this model. Auctioning off cost-sharing contracts in a theoretical and experimental study, highlighting the moral hazard and adverse selection issues in procurement contracting, have been studied by Cox and Isaac (1994). A theoretical model of procurement contracting is developed and tested in laboratory experiments. Market performance is tested for both fixed-price and cost-sharing contracts. Contracts are awarded with first-price sealed-bid or second price sealed-bid auctions. The environment contains postauction cost uncertainty and opportunity for unmonitored effort in contract cost reduction. They find that cost-sharing contracts reduce procurement expense but also are inefficient because of their induced moral hazard waste and cost overruns. This work, although concerned with the cost saving side have similarities and relevance to issues in the patent trade studied here.[39]

[39] This article was brought to my attention by Prof. David Porter, when discussing the two-dimensional contract for patents and previous literature relating to this problem. However, the article does not address two-dimensional bidding of the linear contract but it provides a complete system of models and experimental test.

With direct relation to patent licensing, Kamien and others have done theoretical work on cost-reducing patents (process patents[40]). Kamien has studied patent licensing in a multifaceted way addressing issues of optimal licensing, fees versus royalty in an "either-or-way," private value of a patent, and market structure and innovation (Kamien and Schwartz 1982; Kamien et al. 1992; Kamien 1992). His analysis takes the same starting point as Arrow's analysis from 1962 and develops it toward a market structure using auctions and game theory in joint work with Tauman (Kamien and Tauman 1986, 2002). This game theory work is based in part on the work of Aumann. Aumann (Aumann and Dreze 2009) addresses the question of the management of risk where the "folk theorem" in game theory, with discount rates and punishment strategies, seems to work best. A related study was done by Maskin et al. on repeated games with discounting (compare bidding in auctions) or incomplete information (compare private information in Fudenberg and Maskin 1986). The concept of discounting is directly connected to the cost of capital and management of *financial* risk using the capital asset pricing model (CAPM) (Sharpe 1964), in particular, in its intertemporal version, ICAPM (Merton 1973). The method ties into the investment strategies developed by Markowitz, Miller, and Sharpe especially regarding portfolio selection (Markowitz 1952) and cost of capital (Modigliani and Miller 1958). To conclude this passage, after a one-dimensional contract is auctioned, the risks remain unchanged as before the auction according to theory. This fact that current auction theory does not deal with what happens to the risks after a transaction is consummated was pointed out by Prof. V. Smith during a discussion on spectrum auctions.[41] An importance of the linear contract may be its "ability" to express and negotiate the risk sharing in the contract which is not done through a single fixed price that simply transfers risk.

The equation to be solved by the economic agents is pricing the two market access values *simultaneously*, one has a more "common value" while the other has a more private value. How the market is organized (the messages and rule set of the institution) is the second independent variable in the experiment.[42]

By systematically exploring different messages, information and rules that are commonly used and extensively studied for one-dimensional auction markets both naturally occurring in game theory and in design markets, the attempt is to broaden some of these institutional designs to study prices, adjustment processes, and gains from trade.

---

[40] In general, in the theoretical analysis that has been done with respect to patents and economics, only marginal process patents have been considered, i.e., the study has in general been done on the marginal value of cost-reducing inventions on existing innovations. Here, we are interested in product patents (which create value not only lower cost of existing products) and process patents and the systemic effect, or dynamic effect and marginal effect of patent used in a trade-based economic system.

[41] Refer to presentation made at the Information Economy Project's speaker series on "Big Ideas about Information" at GMU, May 2, 2006.

[42] Refer to Smith (1982) where institutions are made up of messages and rules in an economic analysis. Agents interact with each other *given* the institutional rules to produce outcomes.

Table 2.1 Primary market messages and public information

Institution	Seller's ask		Public information			Buyer's bid
			WTA	WTP		
1	$(f\downarrow, r\downarrow)$/accept	→	(f, r)	(f, r)	←	$(f\uparrow, r\uparrow)$/accept
2	$(f\uparrow, r\downarrow)$/accept	→	(f, r)	$(-, r)$	←	$([f\downarrow], r\uparrow)$/accept
3	$(f\downarrow, r\downarrow)$	→	(f, r)	$(-, -)$	←	Accept/reject

Bidding space: positive and negative values are allowed for (f=fixed fee, r=royalty). Seller starts auctions. Amelioration rules are indicated with ↑ (increase) or ↓ (decrease). Bid=ask or accept stops auctions. In institution 2 ($f\uparrow$) is a de-facto reservation price on [f] asked by the seller

As a first study, different messages are studied for *demand-side* bidding together with different bidding rules to investigate which integration of messages and information are needed to efficiently price the tradable linear contract on patents. In this respect, the message format, information it can contain, and rules under which messages are exchanged characterize the second independent variable. Changes in the execution of the competitive demand-side bidding are achieved by varying the message bid response a buyer can give to a seller's message ask. The bidding takes place in three way conversation: a buyer can send a simple accept/reject as a response to an ask in fixed fee and royalty, negotiate in royalty as a response to an ask (buyers increase royalty), or negotiate in both fixed and royalty to an ask (buyers can increase bids and sellers can decrease asks in one or both dimensions simultaneously). The sell side has the same message in all institutions, i.e., a fixed fee and royalty. The sell side's willingness to accept (WTA) a bid is always public information, whereas information on the demand side's willingness to pay (WTP) changes with the institutional rules. See Table 2.1 for a summary of messages and information in the studied institutions.

The first institution is similar to a two-dimensional version of a double auction except that the seller starts the auction with a first ask. The second institution is a double auction in royalty but with a de facto reservation value on fixed fee (the sellers can increase and buyers decrease the fixed fee which for all practical purposes means that the seller sets a reservation price on fixed fee). The third institution, a posted offer, is similar to a "manual" Dutch clock auction with the seller "ticking" down the asking price in two dimensions independently and simultaneously. The third institution is also similar to today's personal exchange mechanisms, where often a seller exclusively (under NDA agreements) goes from buyer to buyer in a sequential order. The effects of this demand-side bidding with increasing interactivity in the message and public information exchange can therefore be studied systematically for a primary market.[43] These are thus but three examples of many institutional designs that can be studied in the experimental system. The reason for these three is that they have aspects that are commonly

[43] This construction also allows to study the price adjustment process, competitiveness of bidding by varying the number of buyers and robustness of the institutions by varying the induced values.

practiced in the real world and that there have been many experiments run using similar one-dimensional versions (e.g., Coppinger et al. 1980; Cox et al. 1982; Smith 1976; Ketcham et al. 1984; Davis et al. 1990; Easly and Ledyard 1979; Friedman 1984; Friedman and Rust 1993; Friedman 1993; McCabe et al. 1992; Rassenti et al. 1982; Rassenti et al. 2002). Thus, the two-dimensional market designs both draw from the real world and hopefully will add a new dimension to the auction literature as well.

In the secondary market, contracts can be resold in a fixed price double auction, with the fixed fee and royalty price already determined. This market adds to the dynamics of the system as it allows the initially negotiated contract to be tradable for several periods (simulating the patent life). A contract can still have value to another agent, and can be resold to that agent, if the first buyer finds out that the potential earnings from the contract are higher than the price paid. This "exit" possibility also provides some discipline on a buyer whose intention is to Block: getting the contract by simply overbidding on the royalty becomes less sensible. The true investment value would be foregone. The experimental economic system thus allows the study of *dynamic* benefits from the patent system and adds to the literature on *static* benefits.

2.4 Informal Theory of Prices and Initial Hypotheses

This section covers an informal discussion on a price theory for the linear contract and some initial hypotheses that will be tested in the experiments using the dynamic model. Let us first go back to the original argument on values. The economic value of an excludable and transferrable patent right was argued in Sect. 2 to be in the competitive-enhancing technology to "manage" the market access risk related to an innovation. Since the right is excluding, it can be used in a new innovation by making investments in the use of the technology, creating a competitive advantage for own products, or blocking others from doing that and thus limiting competition on current own products to protect the current competitive advantage.

Creating new market access, measured in terms of new sales (revenues), is often a long-term strategy involving considerable uncertainty and major capital expenditure. Maintaining current market access, measured in terms of existing revenues, is on the other hand often a short-term strategy involving minimal capital expenditure and much less uncertainty since current sales are known (and can be forecasted with some certainty). By blocking adversaries, market access can temporarily be maintained by a firm, ceteris paribus, and long-term investment decisions can be delayed or "better timed" to the benefit of cash flows from current assets. The proposition for theory is that blocking to maintain sales that could otherwise be irreversibly lost to a more competitive innovations (own or others) is formally similar to an *insurance* contract, where the risk (calculated with some certainty) of irreversible losses are "traded" for a fixed fee. An identical argument is given by Smith (1968) with respect to inventory being an "insurance" against loss of sales.[44] Smith's work expands on an earlier work by Arrow (1965) on optimal protection of assets against casualty or liability loss and is closely followed in the analysis here. The asset that has to be protected here is the market access (customer relations). The decision that has to be made is different: instead of stocking up products, the strategy is to "blocking" out new competing products. The decision is whether to engage in a "value negotiation" with customers by using the new technology, blocking the new value, and continue to face "price taking" customers until a better technology comes out or loose ground by not buying the technology. The competitive advantage is ultimately given the excluding right attached to the IP in the product. In the IP case, it is the "right" to sell (handed over by the patent right) that provides the "insurance" not the physical availability of the product. These two "insurances" of course work together.

In an insurance contract, a cash payment is paid against the risk of irreversible losses. In the case of the tradable linear contracts on patents, the "insurance" against loss of sales is thus given by the blocking value of that patent right. The proposition here is then that the willingness to pay (WTP) for a patent's blocking value will be expressed in the *fixed* component of the linear contract since this blocking value is the value of "sitting on" the contract and can be owned by any economic agent that could make use of it in an innovation. As with insurance, such risks of loss of market share are pretty well known. Such a value cannot possibly be expressed in a royalty, when the intention is not to use it (which would yield 0 in payments). If competitive bidding is assumed with discrete incremental but uncertain private values, as is tested in the experiments, then under rational expectations the price would be the average of the probability distribution of possible outcomes[45] of the second highest blocking value among buyers, plus epsilon. This case also assumes that the same agent has both the highest blocking and investment values. We can now formulate a first hypothesis:

> H1: The fixed fee that clears the market will be equal to the second highest blocking value among bidders. If Traders are involved, the fee will equal the highest blocking value among Innovators.

By investing in the technology a firm could potentially increase the competitive advantage of its products and "secure" long-term market access but at the cost of a new and uncertain "value negotiation" with the customers. The proposed pricing here is looking at such a decision as a project investment using an NPV calculation of expected (uncertain) sales, discounted with a "project beta." Due to the inherent uncertainty of such projects, many of these projects may be needed to succeed.

---

[44] In Smith 1968, "Optimal Insurance Coverage," p 68, Smith states that "the problem of optimal insurance coverage is formally similar to the problem of optimal inventory stockage under uncertainty. To inventory a product is to "insure" against sales loss – the larger the inventory, given the distribution of demand, the greater the 'insurance coverage'."

[45] In the experiment, a uniform distribution is used and the average is therefore the mid-point of the induced value range.

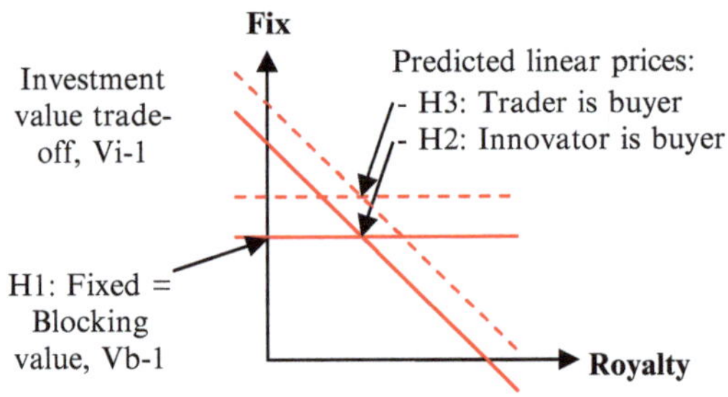

Fig. 2.4 Predicted price in competitive bidding equals the intercept between the blocking value and the investment value linear equations. Prices "above" the lines are unprofitable for their use and "below" profitable for their respective use (Invest/Block). The *dashed lines* are the highest Innovator value and the *solid lines* the second highest value

If the fixed component reflects a blocking value greater than 0, projects could be accepted up to a royalty where the NPV is still positive *after* the cost of the fixed fee (second highest blocking value) has been subtracted. This makes the royalty payment a buyer is willing to pay linearly dependent on the fixed fee offered. See Fig. 2.4 for prediction of prices under rational expectations.

We can now formulate a second hypothesis:

H2: The predicted royalty price (when Innovators are buyers) is then decided by the intercept of the second highest blocking value among agents (since no one would bid higher than that) and consequently the second highest investment value, which is linearly dependent on the fixed fee paid for the blocking value, for an institution that allows simultaneous bidding in both dimensions.

The third hypotheses is related to the dynamics of trade: If a contract is bought by a Trader and split and new contracts are sold for different fields-of-use, the Trader would under rational expectations use the risk sharing of the fixed fee (which requires capital) to outbid the Innovators through the fixed fee. Some of the gains from trade are therefore shared through the market with the Inventor. This would suggest that fixed fees are higher for a contract that is later split. Since Traders always would have the incentive to outbid Innovators, gains from trade would be equal to the multiplicative value given the increase in fields of use.

H3: The fix price would equal the highest blocking value when a Trader is the buyer.

In the case where other induced values are used, for example, the highest blocking and investment values are held by different agents, the prices would not be the second highest but in between the second and the highest value depending on the NPV of the blocking and investment values.[46]

A fourth hypothesis on systemic risk is formulated: The single-firm "invent-and-produce" model and two-firm "invent–produce" model (used in optimal patent

[46] These cases are not presented here but left for future experiments.

licensing studies) are both replaced by the *three*-firm "invent–trade–produce" model in the designed system. This makes the decision problem dynamic with respect to use of technology and funding for development as the contracts can be split by a Trader, potentially multiplying its uses. This may also have effects on the propensity for developing inventions in favor of blocking and lead to a more competitive allocation of capital. In this three-firm trade, capital can thus flow directly into ideas through the contract, independently of the innovation firms, who traditionally fund most of the inventions. (This is a clear trend in many industries, in particular, Pharma-BioTech, where bilateral negotiations make this happen through a business model based on a patents). In the designed system, the increase in the competition for technology makes collusion difficult by traditional Innovators in the presence of the Traders. Instead of investing in stock of an Innovator, Traders/Investors invest in the technology through contracts on patents, split them for different industries/markets/periods, and then sell them to potentially more Innovators. The economic environment and contract designed may allow for *a* new *role* for capital: to fund trade in new ideas (patent portfolio management companies are today becoming a well-established new business model). This creates a *systemic* change in risk sharing among the three parties *during* patent time. The injection of capital in this trade to bear risks better in the system (a risk absorption mechanism facilitated by the market) creates the possibility of dynamic gains from trade and competitive decision implemented by this market rather than a hierarchical structure. With the additional Trader bidding for contracts, the risk is allocated more efficiently, potentially lowering the systemic risk. Competitive bidding may not only attract capital to trade the ideas, but also to develop them from (for example) the venture capital industry. This suggests that it will not be as easy to Block good ideas when they can be competitively priced and gains extracted from potentially multiple innovations. The conclusion here suggest that there is a reduction of cost of capital in the system due to more efficient allocation of risk bearing by the actor with best capability of managing risks given the specialized nature of his firm.

H4: A market with transparent prices would lead to a more efficient economic system with respect to use of technology enabled by a more efficient use of capital.

The last proposition is that the price system itself signals the value of a technology, potentially directing the technology investment focus (technical area) of an Inventor to an area of *commercial* interest.

H5: More economically useful inventions would be developed when coordination takes place through a market with prices.

These propositions are contrary to the conclusions of Robinson, who in her article "What are the Questions?" (1977, p. 1337), suggested the problem in the economy to be that the capital to fund new innovations was accumulated by large corporations and any development decision was at their discretion (emphasis added):

> Here is the problem. The task of deciding how resources should be allocated is not fulfilled by the market but the great corporations who are in charge of the *finance for development*. These questions involve the whole political and social system of the capitalist world; …

Competitive trade in technical knowledge appears to create competition for capital between Inventors and Innovators, opening for directly funding a market in technology.

2.5 Summary

In this paper, I have proposed a dynamic economic system design for trading patent rights in organized impersonal markets with prices useful for multiple experimental works. An informal theory of pricing a linear contract on the patents has been outlined and initial hypotheses formulated. Two initial experiments have been proposed for initial study: (a) linear prices and gains from trade investigating performance and behavioral characteristics for different design mechanisms and legal environments, and (b) coordination of inventive, investment, and innovative activities with market prices as price signals, for the same mechanisms and changes in environment.

The model allows flexibility in several of the parameterized dimensions for these and future studies on markets, of which the following appears the most important contribution toward a more dynamic analysis of the patent system as a trade system:

1. Multiple mechanism designs. Different auction mechanism designs can be tested. The information content of the messages can be varied as well as what private information is accessible publicly through the institution. The adjustment process can be studied to evaluate endogenous bidding strategies. The incentive schemes (of rules) can be studied.
2. A dynamic patent system. Different presumed validities can be used and validity can change over time (between experimental periods). Also information on patent dimensions can be varied by field-of-use, and in-validation of patents can occur in markets and time.
3. The contract can be of exclusive or nonexclusive type. Different number of licenses of the patent can be used in studying prices and coordination. Special types of contracts such as "bundling" of patent rights can be included.
4. The induced values can vary over time. Technology of different lifetimes can be studied. The economic value of a technology varies considerably, from months to decades. The model allows varying values over time to simulate typical technology patterns.
5. The uncertainty in the induced values can vary. This uncertainty can be varied through different value distributions, ranges, and averages of the induced values.
6. Additional risk mitigation modes can be varied. Insurance against loss of validity and other "insurable" risks can be included to study the effect on prices.
7. The induced values can be varied. Randomizing values of technology allows effective study of n-commodity pricing under rational expectations. Convergence

to rational expectations can be studied.[47] Also negative values can be used to study the limits of voluntary trade and moral hazards.

8. The competition environment can be varied. The number of Inventors, Traders, and Innovators can be varied to study situations with few and many competitors. This is particularly of interest since there are often a limited number of firms who can use a certain technology.
9. The "technology" environment can be varied. Different "search limitations" can be designed to study coordination between invention, trade, and innovation as well as risk-taking behavior during the search.

The proposed initial two experiments, reported in this book in Chaps. 3 and 4, thus aim at studying integration of messages, information and institutional rules under varied validity and resulting dynamic[48] prices, and gains from trade amounting to two controlled experimental economic laboratory studies of the performance and behavioral properties of the linear contract and mechanism designs for organized markets in patents.

References

Arrow, J. K. (1962). Economic Welfare and the Allocation of Resources for Invention. *In NBER The Rate and Direction of Inventive Activity: Economic and Social Factors. Princeton Univ. Press (for the NBER).*

Arrow, K. J. (1965). *Aspects of the theory of risk-bearing*. Helsinki: Yrjö Jahnssonin Säätiö.

Aumann, R. J. & Dreze, J. H. (2009). Assessing strategic risk. *American Economic Journal: Microeconomics, 1*(1), 1–16.

Coase, R. H. (1960). The problem of social cost. *Journal of Law and Economics, 3*, 1–44.

Coppinger, V. M., Smith, V. L., & Titus, J. A. (1980). Incentives and behavior in English, Dutch and sealed-bid auctions. *Economic Inquiry, 18*, 1–22.

Cox, J. C., Roberson, B. & Smith, V. L. (1982). Theory and Behavior of Single Object Auctions. *Research in Experimental Economics*, 2(Greenwich: JAI Press).

Cox, J., Isaac, M. R., Cech, P.-A., & Conn, D. (1994). Moral hazard and adverse selection in procurement contracting. *Games and Economic Behavior, 17*, 147–176.

Davis, D., Holt, C. A., & Villamil, A. P. (1990). Supracompetitive prices and market power in posted-offer experiments. University of Illinois BBER Faculty Working Paper No. 90–1648.

Easly, D., & Ledyard, J. (1979). Simultaneous double oral auction markets. Public Choice Society Meetings, March 1979.

England (1623). English statute of monopolies of 1623, 21 Jac. 1, c. 3, The Original Source of the Anglo-American Patent Law.

Friedman, D. (1984). On the efficiency of experimental double auction markets. *The American Economic Review, 74*, 60–72.

[47] Smith makes the point that rational expectations with respect to sustainability (Nash) or theory (Muth) can be observed, making it possible to study expectations that are not supported by the theory of Smith et al. (1988).

[48] Since the dynamic system may have different end states (like in dynamic programming), realized experimental gains could be divided by one or more optimal end states to create a measure capturing some of the dynamics. Dynamic efficiency = realized gains/optimal end state.

Friedman, D. (1993). The Double Auction Market Institution: A Survey. In *The Double Auction Market: Institutions, Theories, and Evidence*. Addison Wesley Publishing Company, pp. 3–25.

Friedman, D., & Rust, J. (1993). *The double auction market: Institutions, theories, and evidence*. New York: Addison Wesley Publishing Company.

Fudenberg, D., & Maskin, E. (1986). The folk theorem in repeated games with discounting or with incomplete information. *Econometrica, 54*, 533–554.

Gambardella, A., Giuri, P., and Mariani, M. (2005). The value of European patents evidence from a survey of European Inventors. Final Report of the PATVAL EU project, Contract HPV2-CT-2001-00013, January 2005.

Grief, A. (2002). Institutions and Impersonal Exchange. *Journal of Institutional and Theoretical Economics, 158*, 168–204.

Hart, O. D. (1988). Incomplete contracts and the theory of the firm. *Journal of Law, Economics, & Organization, 4*, 119–139.

Hart, O., & Moore, J. (1988). Incomplete contracts and renegotiation. *Econometrica, 56*, 755–785.

Kamien, M. I. (1992). Patent Licensing. *Handbook of Game Theory with Economic Applications (book),* 1(Chapter 11), pp. 331–354. Elsevier BV.

Kamien, M. I., Oren, S. S., & Tauman, Y. (1992). Optimal licensing of cost-reducing innovation. *Journal of Mathematical Economics, 21*, 483–508.

Kamien, M. I., & Schwartz, N. L. (1982). *Market structure and innovation*. Cambridge: Cambridge University Press.

Kamien, M. I., & Tauman, Y. (1986). Fees versus royalties and the private value of a patent. *The Quarterly Journal of Economics, 101*, 471–492.

Kamien, M. I., & Tauman, Y. (2002). Patent licensing: The inside story. *The Manchester School, 70*, 7–15.

Katz, M. L., & Shapiro, C. (1985). On the licensing of innovations. *The RAND Journal of Economics, 16*, 504–520.

Ketcham, J., Smith, V. L., & Williams, A. W. (1984). A comparison of posted-offer and double-auction pricing institutions. *The Review of Economic Studies, 51*, 595–614.

Knight, F. H. (1921). *Risk, uncertainty and profit*. New York: AM Kelley.

Krugman, P. R. (1990). *Rethinking international trade*. Cambridge: MIT Press.

Laffont, J.-J., & Tirole, J. (1987). Auctioning incentive contracts. *The Journal of Political Economy, 95*, 921–937.

Lamoreaux, N. R., & Sokoloff, K. L. (1999). Inventive Activity and the Market for Technology in the United States, 1840–1920. NBER Working Paper No. 7107.

Lamoreaux, N. R., & Sokoloff, K. L. (2001). Market trade in patents and the rise of a class of specialized inventors in the 19th-century United States. *The American Economic Review, 91(2),* Papers and Proceedings of the Hundred Thirteenth Annual Meeting of the American Economic Association, 39–44.

Markowitz, H. (1952). Portfolio selection. *The Journal of Finance, 7*, 77–91.

Maskin, E., & Riley, J. (1984). Optimal auctions with risk averse buyers. *Econometrica, 52*, 1473–1518.

McCabe, K. A., Rassenti, S. J., & Smith, V. L. (1992). Designing a uniform-price double auction: An experimental evaluation. In D. Friedman & J. Rust (Eds.), *The double auction market: Institutions, theories and evidence* (pp. 307–332). Reading, MA: Addison Wesley.

Merton, R. C. (1973). An intertemporal capital asset pricing model. *Econometrica, 41*, 867–887.

Milgrom, P. R., & Weber, R. J. (1982). A theory of auctions and competitive bidding. *Econometrica, 50*, 1089–1122.

Modigliani, F., & Miller, M. H. (1958). The cost of capital, corporation finance and the theory of investment. *The American Economic Review, 48*, 261–297.

Plant, A. (1934). The economic theory concerning patents for inventions. *Economica, 1*, 30–51.

Porter, D., Rassenti, S., Roopnarine, A., & Smith, V. (2003). Combinatorial auction design. *Proceedings of the National Academy of Science of the United States of America, 100*, 11153–11157.

Porter, D., et al. (2009). The Design, Testing, and Implementation of Virginia's NOx Allowance Auction. *Journal of Economic Behavior and Organization, 69*(2), 190–200.

Radner, R. (1995). Economic survival. Nancy L. Scwartz Lecture, Kellogg School of Management, Northwestern University.

Rassenti, S., Smith, V. L., & Bulfin, R. (1982). A combinatorial mechanism for airport time slot allocation. *Bell Journal of Economics, 13*, 402–417.

Rassenti, J. S., Smith, L. V., & Wilson, J. B. (2002). Demand-side bidding will reduce the level and volatility of electricity prices. *The Independent Review, 6*(3), 441–445.

Razgaitis, R. (2006). U.S./Canadian Licensing In 2005; Survey Results. *LES Nouvelles.*

Razgaitis, R. (2007). U.S./Canadian Licensing In 2006; Survey Results. *LES Nouvelles.*

Robinson, J. (1977). What are the questions? *Journal of Economic Literature, 15*, 1318–1339.

Samuelson, W. (1986). Bidding for contracts. *Management Science, 32*, 1533–1550.

Samuelson, W. (1987). Auctions with contingent payments: Comment. *The American Economic Review, 77*, 740–745.

Schumpeter, J. A. (1942). *Capitalism, Socialism and Democracy*, New York: Harper.

Sharpe, W. F. (1964). Capital asset prices: A theory of market equilibrium under conditions of risk. *The Journal of Finance, 19*, 425–442.

Smith, V. L. (1968). Optimal insurance coverage. *The Journal of Political Economy, 76*, 68–77.

Smith, V. L. (1976). Bidding and auctioning institutions: Experimental results. In Y. Amihud (Ed.), *Bidding and auctioning for procurement and allocation* (pp. 43–64). New York: New York University.

Smith, V. L. (1982). Microeconomic systems as an experimental science. *The American Economic Review, 72*, 923–955.

Smith, V. L., Suchanek, G. L., & Williams, A. W. (1988). Bubbles, crashes, and endogenous expectations in experimental spot asset markets. *Econometrica: Journal of the Econometric Society, 56*, 1119–1151.

Smith, V. L. (2003). Markets, Institutions and Experiments. *Encyclopedia of Cognitive Science*, L. Nadel (ed.), Macmillan Publishers (Nature Publishing Group).

Smith, V. L. (2007). *Rationality in Economics: Constructivist and Ecological Forms*, Cambridge University Press.

Tirole, J. (1999). Incomplete contracts: Where do we stand? *Econometrica, 67*, 741–781.

Ullberg, E. (2009). From personal to impersonal exchange in ideas – Experimental study of trade in organized markets for patents. *KTH TRITA-TEC-PHD 09–006*,180.

Vickrey, W. (1962). Auctions and bidding games. In *Recent advances in game theory*. Princeton, NJ: Princeton University Press.

Vickrey, W. (1976). Auctions, markets, and optimal allocation. In Y. Amihud (Ed.), *Bidding and auctioning for procurement and allocation*. New York: New York University.

WIPO. (1883). The Paris Convention for the protection of industrial property. *WIPO Database of Intellectual Property Legislative Texts.*

WTO. (1994). *The TRIPS Agreement*, Available at: http://www.wto.org/english/tratop_e/trips_e/trips_e.htm.

Chapter 3
Prices and Dynamic Gains in Markets for Patents: Experimental Study of Impersonal Exchange in Ideas

3.1 Introduction[1]

In this chapter, I wish to examine trade in patents in their own rights in organized markets with transparent prices. The economy has always been an "intellectual property" economy based on specialized agents trading products and services, using different mechanisms to keep the knowledge needed to produce private, but through the patent system knowledge[2] has become tradable.[3] The patent system changes the structure of economic organization by potentially introducing a competitive market in technology through the transferrable and licensable right on the technical ideas themselves. Such a market creates a dynamic economic system where increases in the stock of technology and focus for technology (technical area) are a result of competitive demand-side bidding, as bidders express their value for the technology resulting in transparent prices. Such shift in the focus of economic organization can be expressed as a shift from a personal to an impersonal exchange in ideas, a process that got started with the first patent system in 1474. This institutional change in the economic system made private technology both (a) public through publication of the (disclosed) technology (a social exchange and gain) and (b) tradable (a market exchange and gain) through an excluding, and transferrable and licensable right

---

[1] The chapter, primarily concerned with the dynamic outcomes of trading patent rights in an experimental economic system, arose out of the study of management of risk and uncertainty through new market mechanism designs at ICES-GMU, and is part of a broader research on markets in IPR or "markets in ideas." See Ullberg (2009). Special thanks are expressed to the Savings Banks Research Foundation, Dir. Rodriguez for funding this research, The Royal Institute of Technology in Stockholm (KTH), and numerous personal communications with Prof. V. Smith, S. Rassenti, D. Porter, O. Hart, as well as input on market design from Goldman Sachs.

[2] Technichal ideas and, also nontechnical ideas in the USA, can be patent protected for products and processes.

[3] The patent systems typically give the owner two rights: to exclude and to transfer or license the right. It is the second right, to transfer or license, that is a focus in this study.

E. Ullberg, *Trade in Ideas: Performance and Behavioral Properties of Markets in Patents*, Innovation, Technology, and Knowledge Management 13,
DOI 10.1007/978-1-4614-1272-4_3, © Springer Science+Business Media, LLC 2012

on new technology.[4] In this chapter, the focus will be the "last step" in this transition toward impersonal exchange and three design mechanisms used for exchanging licensing contracts on patents.

An experimental and dynamic economic system design was developed in an earlier article (Ullberg 2010c) and in Chap. 2. This model, developed for a broad range of *market exchange and social exchange* studies on patents, is used here with an experimental design to begin to investigate some of the questions central to impersonal exchange: transparent prices of license contracts on patents (given different institutional arrangements, investigating the proper integration of information and rules) and dynamic gains from exchange (in terms of maximizing the use of technology in the economic system). This is done in two experimental studies. This first study, reported in this Chapter, is a joint study of the traded contract and *market* mechanisms for trading patents in organized impersonal markets with transparent prices in a dynamic economic system, where the patent validity is varied ("high" and "low" validity). This results in a 3×2 experimental design. The *social* exchange in the economic system is explored in a setting of a broader coordination of demand for technology and investment with certain technology focus in a second experiment building on the first experiment. Those results are reported in the next chapter and in Ullberg (2010a).

The key question addressed in the experimental design are (a) mechanism designs (whose price outcomes are compared to that of the proposed price theory), (b) a heuristic analysis where different designs are compared with respect to differences in prices and what can be learned from their behavioral properties (giving information on proper integration of information and rules for this kind of market and social exchange), (c) dynamic gains given the designs and patent validity (the dynamic outcome/optimal social dynamic outcome), and (d) change in risk in the economic system (measured as a calculation of cost of capital in the system as activities are coordinated through prices).

Before discussing the trading system, I state briefly the principal findings from the first experiment: (a) The (linear) contract prices appear to shift the risk-bearing away from the Inventor to the Innovator (and as a result ultimately to the consumer) in a way that supports the proposed informal price theory (best in an institution with two-dimensional bidding), (b) reducing the risk in inventive activity as such which is likely to increases the competition in technology, and (c) the dynamic efficiency (the use of technology in new innovations) is at least doubled by introducing demand-side bidding on contracts for patents.

A short summary of the model (trade system), prices theory, and approach (mechanism designs) described in Chap. 2 is first done as a background to the experiment (Sect. 3.1). For a more detailed explanation of these considerations, see (Ullberg 2010c). This is followed by a description of the implemented dynamic

[4] Typical criterias of patentability are that the technology has to be: new, nonobvious/have a sufficient inventive step, and be useful/have industrial applicability. This excludes scientific formulas and the most part nontechnology (with exception of business processes in the USA).

microeconomic system model for the experiments (Sect. 3.2), the experimental results, and some initial hypotheses (Setcs. 3.3–3.4). Conclusions on what has been learned about prices and dynamic gains, with possible policy and future research implications, concludes this Chapter (Sect. 3.5). Some of the policy implications presented are further elaborated in a separate note included in a report entitled "Patents markets in the knowledge economy" for the economic counsel of the French prime minister (Ullberg 2010b). This note is included in the epilogue.

3.1.1 The Patent System as a Trade System

The economic system design is based on the *principles* of the patent system in order to make conclusions more relevant for policy. The problem of trading patents is here focused on *impersonal* exchange, i.e., a mechanism design problem, for a linear contract, in an environment with dynamic gains (multiple use of technology), uncertain technology values, and only presumed legal validity of the patent (uncertain patent rights). Two questions arise: how the prices should be determined on patents in a dynamic environment (the market studied is thus a producer market for technology) and how dynamic gains from trade may contribute to the value of the patent system. Impersonal exchange markets for patents are unknown; no formal exchange exists to date,[5] and it is almost impossible to get reasonable information about how such a market could work (most personal exchange contracts are private with private prices). Therefore, an experimental economic approach is chosen as a method of study.

The patent system is used in a complex manner. It is currently an almost global system but it is in nature national/regional, with fundamental importance to national economies and trade agreements. It is increasingly used by an ever-broadening range of individuals, firms, and nation states,[6] and also encompasses an increasing range of patentable subject matter, which in recent decades has expanded (to some extent) to include nontechnical areas such as financial and internet-based services. This broad use makes it "inseparable" from economic activity today.[7] The approach chosen in this experiment to deal with the complexity is to take the *principles* on which the patent system is built as a basis for an economic system design and study one central aspect: the trade aspect, i.e., prices and possible dynamic gains from trade. A patent system relies on few, but powerful, legal principles: public disclosure

---

[5] An almost explosive development in patent licensing has been observed over the last 25 years, after the 1982 change of the US patent system. A similar trade in patents was observed after the first US patent law in 1832 (Lamoreaux and Sokoloff 2001). Recently, fixed-price auctions of patents have successfully been run by some private specialiced firms.

[6] Local inventors, university professors, inventor companies, governments and banks/venture capitalists as collateral for debt, etc.

[7] Similar comments can be made for other intellectual property rights covering other areas like: trademarks, copyrights, geograhpical indications, etc. (WTO/TRIPS have currently seven types).

of what is invented, the private right to exclude, transfer, or license, the priority date for the invention, and national treatment, i.e., nondiscriminatory treatment of international inventors.[8]

This development of property rights in the history of technology has also affected the organization of economic activity related to inventions, changing it from a single hierarchy *toward* coordination between agents in a market with prices. Specialized inventive firms have appeared, trading their ideas through personal exchange of patent rights. An early example of this was already observed after the 1832 patent reform in the USA, where a market for technology developed for some time (Lamoreaux and Sokoloff 1999). Since the 1982 decision, a high growth in patent licensing has been observed, especially in new technologies, which reaches perhaps one trillion dollar annually[9] with different kinds of specialized firms, from pure inventing to joint ventures. Likewise, new types of agents have developed, such as patent portfolio holders.[10] Most of these trades take place as a personal (and bilateral) exchange between firms with a few exceptions of initial impersonal exchange through public auctions (multilateral bankruptcy auctions and specialized auctions).

The principle used in the experiment is coordination of agents performing inventive, trading, and innovating activities through a market with prices. The division of these agent characteristics is motivated by their respective riskiness, as expressed by the cost of capital.

3.1.2 Summary of a Price System on Patents

This summary is based on the previous model article. There are two problems to be solved by a pricing system: (1) how much money each agent shall have and (2) the system of prices (mechanism) in accordance with which the contracts are to be made available to the buyers.[11] In this experiment, the first question is addressed with an endowed discount rate (cost of capital) to calculate the value of a contract; and a role-specific exchange rate of experimental dollars to paid dollars intended to, in principle, make each role equally profitable.[12] The second question is the main question of investigation.

[8] There are also some exeptions related to risks for social unrest, which in fact are caveats of the system.

[9] This is an estimate made by many patent lawyers.

[10] This business model opens the door for a new use of capital-to-trade technology. These agents bridge the time between invention and innovation and capitalize on the maximum use of technology. This business model is thus incentive compatible with social gains. Such a use of capital is incorporated in the experiment as "traders" who take positions in a contract, split them, and sell two new contracts.

[11] These criterias of a pricing system is made with reference to Coase (1946).

[12] The motivation behind this is that subjects be equally cash incentivized in any role in the experiment.

First, the mechanisms represent the terms under which an agent can obtain a contract. For fixed-price contracts, "the same factor (contract) should have the same price in whatever use it is employed since otherwise customers (agents) would not be able to choose rationally, on the basis of price, the use in which they prefer a factor (contract) to be employed." (Coase 1946) (words in parentheses added). However, in the case of the patent, it has *two* values depending on its strategic use as discussed in the model article: an investment value and a blocking value. The values differ considerably with respect to (market access) risk. Investing is a high-cost, high-uncertainty activity that provides the holder with a possible long-term competitive advantage, and, therefore, long-term market access from *new* products and processes. Blocking, i.e., "sitting on" the contract, is typically a rather low-cost, low-uncertainty activity that provides the holder with an "insurance" against short-term loss of competitive advantage and market access to *existing* products and processes based on other existing technology held by the firm. These values clearly have to be dealt with separately in a pricing system where rational choices could be made with respect to its use, since the price of a fixed-price contract otherwise would yield different prices depending on its use.

Blocking allows the holder to block competition (temporarily) or to better time its own investment decisions, in order to avoid cannibalization of its own market share, avoid retiring productive assets, or make other possible strategic decisions with respect to research. In the referred article, the proposition is made that blocking is formally similar to an insurance contract [in this case, insurance against irreversible loss of sale (revenues)] and could be *priced* as such. This suggests that the blocking value could be priced with a fixed price, since the risk is more or less predictable and therefore "insurable" and transferrable.

On the other hand, the high uncertainty of the investment value is typically not predictable in any way similar to blocking. The risks are not "insurable," nor transferrable (since it is hard to put a value on them ex ante their use in an innovation). The proposed solution here is a risk-sharing arrangement based on a royalty. We here borrow from investment theory and propose that the net present value (NPV) of the investment value *minus* the paid fixed price (attributed to the blocking value) would be paid as long as it is positive. A royalty on sales (revenues) would here be possible, based on the realized usage of the contract, thus a price is paid when the risks have become known. This makes the royalty an "investment option" type of arrangement.

These propositions to price the blocking and investment values differing with respect to risk, thus lead to a linear contract with a fixed fee and a royalty component. The proposed contract leads to a multipart pricing system. The characteristics of a pricing system for patents (legally defined as "chose in action") thus sharply differ from those of potatoes and shoes ("chose in possession"). This type of contract is a contract commonly used in patent licensing and preferred among Inventors, although it appears difficult to obtain a fixed component in existing personal exchange environments.[13] Ideally, these two values should therefore be priced

[13] Data obtained courtesy of RoyaltyStat LLC, which holds a database of 8,000+ licensing contracts with 3,000 patent licensing contracts, indicates that only 50% of contracts have a fixed component. Twelve percentage have milestone payments, 24% a lumpsum payment, and 13% a minimum payment.

independently in an auction mechanism. Therefore, coordination between specialized agents should, in order to capture both values, also be able to price these values independently. The proper integration of information and rules will be investigated in three different mechanism designs in this experiment.

Second, a criterion for a system of prices is that the price should be the same for all agents. This means that the *highest bidder* would get the contract. For a traded commodity, this means that supply equals demand, i.e., the price is an equilibrium price. For an exclusive contract with discrete values, of which there is only one available, this means that the expected price should equal the *second-highest* value plus epsilon (Bertrand competition). Since the blocking value can be "immediately" realized by whoever is sitting on the contract (it can be sold to a "blocker"), it is a more "common" value[14] than the investment value. Following this reasoning, the market access value for blocking must therefore be paid by the investing User as well. The consequence is that with the proposition for a theoretical "equilibrium price" for the contract on a patent where the fixed price then equals the second-highest blocking value plus epsilon and the royalty price equals a royalty that keeps the NPV positive, at the given discount rate, for the second-highest investment value plus epsilon, minus the blocking value.[15] The prediction is under the condition that the *same* agent has both the highest blocking and investment values. If these are different agents then the prices will be *between* the highest and the second-highest values in a combination that depends on the size of the blocking and investment values in NPV terms. These "mixed" price predictions are not investigated in this experiment.

The clearing of a linear contract thus is not a simple transfer of an asset like in the typical single-price auction market, but an allocation of risk-bearing and -sharing negotiated *in* the fixed fee plus royalty price. This differs from the traditional analysis of asset pricing where the risk-bearing is separated from a transfer price through derivates on the state of nature (Arrow 1962a) (Arrow-Debreu securities). The two-dimensional price is tested in three market mechanisms.

3.1.3 The Mechanisms Design Criteria

The division of characteristics of the agents has been done with respect to these activities' typical riskiness and mimics the activities typically coordinated in a firm hierarchy by a general manager (research, finance and market, with the general manager being replaced by the patent market with prices). The focus of the design

---

[14] The meaning of "common value" may need to be clarified further for the patent case. Here it is used with reference to a value that is commonly expected among all agents through experience with similar technologies.

[15] In a situation where there are several inventors and competing technologies for sale, a multicontract market in ideas, an equilibrium price would form where the risk in the different technologies is taken into account.

is in the *during*-patent time trade and use of privately owned but publicly disclosed technology (pricing of the linear contract and dynamic gains from trade), not the *pre*patent time private research (e.g., patent races) or *post*patent (e.g., spill-over) use of public technology. The economic environment is the controlled environment, and the institutions and legal environment are the independent variables in this experiment. Values for the contracts are induced and uncertain (given in a range) and different agents have different values with one having the highest value (to investigate the allocation mechanisms). The type of patents examined are product patents, adding revenues by adding "features" to existing products, in contrast to the small cost-reducing process patents often used in the traditional analysis.

The differences between the institutions are the messages and information, as well as the rules under which a *buyer* can respond to a seller. The buyers face the decision to use either both (fixed fee plus royalty), one (royalty), or none (accept/reject) of the linear dimensions to negotiate a price. Thus, they are more or less constrained in expressing their willingness to pay (WTP) for the two values. In the first case, they can fully express their WTP. In the second, for all practical purposes only the royalty can be used. Finally, in the third case, they can only accept or reject an offer posted by the sellers. The sellers can always express their willingness to accept (WTA) using both dimensions. The third institution is close to personal exchange and the others more impersonal exchange. Differences in prices and dynamic efficiency here would suggest possible gains in the economic system given a shift in trade toward a more impersonal exchange system for patents.

These different mechanisms are thought to be useful in studying integration of information/messages and rules for demand-side bidding mechanisms. For the purpose of this study, the mechanisms should reflect agents' pricing behavior in solving the two-dimensional value problem faced (Block/Invest) using a *linear* contract to price them (fixed fee plus royalty). The behavioral question is therefore how the human subjects would express the two values in the two-dimensional price. The hypothesis is that they would do this according to the different risk inherent in the two values.

3.2 The Mechanisms: Details of the Institutions, the Economic and Legal Environments

The trading procedure employed in this study is a specially built computerized trading system[16] for linear contracts that incorporates the three institutional mechanisms tested (the primary markets), individual screens for the Inventor (Role 1), a Trader

[16] The software was developed by the author, except the network messaging module used at ICES for many years. A significant part of the total time in this project was dedicated to developing the trading environment. Special thanks to Lance, a software consultant to ICES, who came up with a key concept to solve the complex endogenous experimental sequence (a state machine) at a critical time in the project.

(Role 2), and the two types of Innovator roles (Role 3A and Role 3B), as well as a fixed-price double-auction mechanism (the secondary market) used to re-trade contracts in periods 2 and 3, already negotiated in period 1 (in the primary market). There are thus three periods in each round. Appendix A provides the participant's screen displays for different roles. An instruction-set used during the experiment allowed participants to learn the interface more quickly by providing each participant with a detailed explanation of different areas, boxes, and information on their screen and what each role could do.

3.2.1 The Endogenous Experimental Flow

Figure 3.1 gives an overview of the state diagram of the endogenous experimental flow in the experiment. The experimental flow follows three general "phases" or steps which are repeated in each round: Inventive step (1), Trading steps (2) and (3), and Using steps (6), (4), and (5). The actual flow is then executed *by the decisions of the participants*. The Inventor is first asked to create a linear contract of "standard" or "quality" type (1). The contract is predefined as a license on an invention with "technology focus" "AB," useful for producing products of types "A" and "B." A quality contract can be split into one contract with focus "A" only and another contract with focus "B" only. The Trader is the agent who can split the AB contract into A and B; thus, a quality contract allows the Trader to participate in the bidding process (2). If the Inventor decides to invest in a standard contract, which cannot be split, the Trader is left out of the bidding (2) for that round. If the Trader wins the AB contract, and then subsequently splits it, the A and B contracts are sold in sequence to the Users of Role 3A and Role 3B, respectively, with the Trader now being the seller and the Users the buyers (3). A User of type A can only produce products of type A (their "product mix") and vice versa for type B Users. If a standard contract is sold, all Users can thus participate but there will be only one winner and one participant (A *or* B type) who can use the contract. If the contract is split, there are two participants (A *and* B type) who can use the two contracts.

The contract(s) sold thus eventually end up in the "portfolio" of one User (with product mix A or B), or two Users (one with product mix A and one with product mix B) and a Trader (who holds the AB contract issued by the Inventor). The User(s) holding the contract(s) are now asked (6) to either "Invest," which produces a profit based on increase in sales of the new more competitive product(s) A and/or B invested in, or to "Block," which produces a profit based on the existing sales of the product(s) based on existing technology the firm has been endowed with, "insuring" the firms market access against competition based on technology. The sales and blocking profits are randomly drawn from a uniform distribution from the holders induced value range and displayed to the participant if "Invest" is chosen (since the issuer should know the basis for the royalties), after which period earnings are calculated for period 1 and the experiment moves to the next period (7).

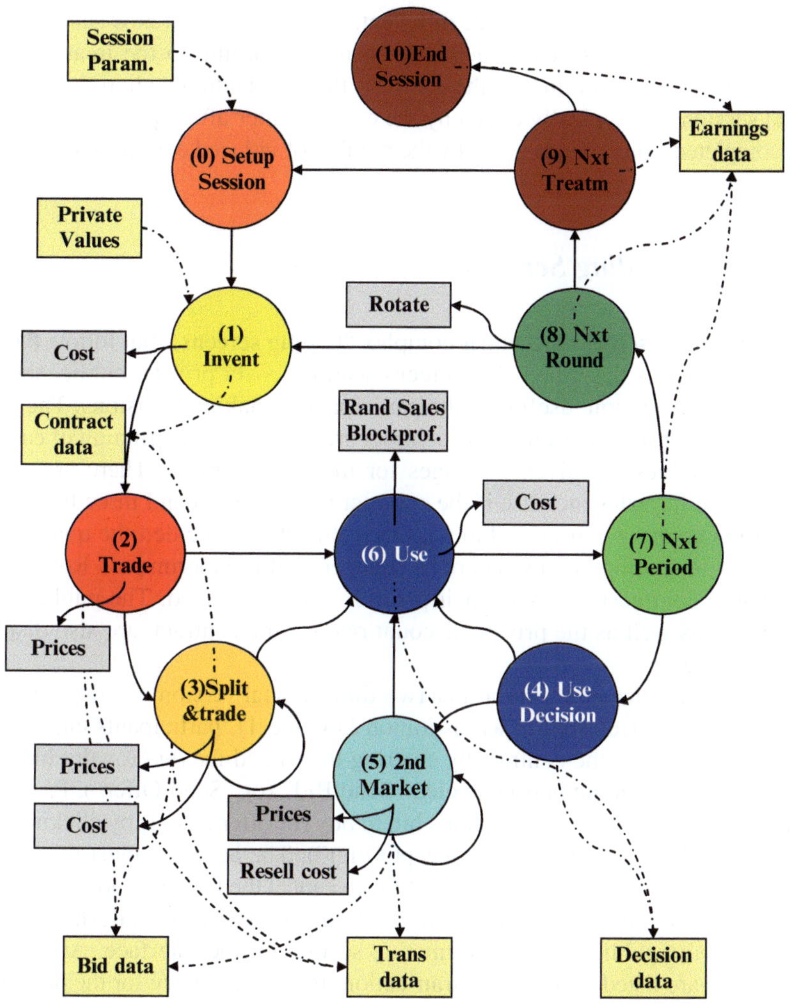

Fig. 3.1 State diagram of the endogenous experimental flow in the experiment

At the beginning of periods 2 and 3, the holder (4) is given the possibility to decide to resell the contract (or keep the contract), which starts the secondary market (5); then, if the contract is resold, the new holder is asked to use it (6) (cannot resell until next period). If the holder decides to keep the contract, then the holder is asked how to use it in this period (6), after which the sales/blocking profits are randomly drawn anew within the respective value range, and period earnings are calculated for all participants who held (bought/resold) or issued (sold) a contract, ending period 2. Period 3 proceeds in the same manner as period 2. At the end of period 3, total earnings are calculated for the round and the next round is started (8). In this round, the roles may be rotated among participants with some frequency (e.g., every two or three rounds). The sequence of Invent, Trade, and Invest now starts again for another round. The induced values may change according to the treatment used. Typically there are two values of the contracts that are randomly altered. The experimental session was typically ended by the experimenter after up to 30 rounds have been completed (max rounds given by the number of induced contracts) or about 2 h.

3.2.2 The Trading Screens

The participant's screens are rather complex "trading screens," but follow the same theme and logic as described. The screens sections cover private, public, and earnings/status information, useful for the participant in making decisions. The private values and other information on the contract are given at the beginning of each new round. The values are given in ranges for the three periods. There are thus six ranges (three for sales increase if the contract is invested in and three for blocking profit if the contract is used to block). Once Role 1 has decided the quality of the contract[17] – technology focus cannot be changed in this experiment – Roles 3 and 2 (if a quality contract) will have their private values displayed. The public validity and quality, as well as the private discount rate for the contract, are also displayed with the values.

In the first mechanism, similar to a two-dimensional version of a double-auction with free bidding (referred to as institution DA-rule 1), participants enter a linear price to buy (or sell) the contract by entering their fixed fee *and* royalty bid (offer) and then clicking on the button labeled "Send Bid" (or "Send Offer"). Participants can also accept any other participant's bid to buy (or offer to sell) by clicking on the button labeled "Accept Bid" (or "Accept Offer"). The acceptor must then confirm the acceptance by clicking "YES" (or "NO" to cancel the bid) on a pop-up box. This results in a binding contract being formed and the exchange information being recorded in the public market information section on the interface. Also, all past transactions are listed with the last transaction first as a memory for the participants regarding negotiated prices for different markets (primary market AB, A, B, resell price, validity, and quality). The participants are given calculation tools to automatically calculate expected profits, given their expectations of the outcome in the value range and whether investing or blocking is chosen. A "sliding ruler" – scaled from 0 to 100% – is used to input values from the included *ranges*, periods (1–2–3), and uses (Invest/Block) into the calculator. The selected *expected* values are displayed below the value ranges and next to the entered quote and calculated expected profits. Expected profits are calculated in real time to allow the participants to test different

[17] In the reported experiment, the inventor's technology focus ("AB") and the user's product mix ("A" or "B") are fixed; only tradability (quality) of contract can be decided. In a second experiment, the technology focus or the contract can be engogenously searched from a limited "search space."

expectations and combinations of fixed and royalty bids *before* submitting their quote.[18] The arrangement allows for studying the risk behavior of the subjects (risk averse, risk neutral, or risk taking).

Price quotes must reduce the bid–ask spread in one or both dimensions in order to be accepted. A buyer can ameliorate a bid, for example, by increasing the fixed bid without increasing the royalty bid. The five highest bids and lowest asks are displayed in a ranked order visible to all participants. Only the highest bid and the lowest ask are open for acceptance. Price quotes that violate this rule are rejected. Identical bids are thus rejected (since there is only one license for sale). The auction is always started by the seller (a Role 1 or Role 2) and ends when the bid and ask meet in both dimensions, an "accept" is made, or the auction is timed out, which results in a "no trade" for that contract. In the case of "no trade," the seller has to bear any costs associated with the creation of the contract (Role 1) or loss of fixed and royalty fees to cover cost against obligations from the contract bought (Role 2). Buyers are not affected by a "no sale." The timeout function has two timers to speed up the bidding process: one for the maximum total time for the auction (usually 120 s) and one for the maximum time between bids (usually 20 s).

The second mechanism, double-auction with reservation value on the fixed component (DA-rule 2), is identical to DA1, except with regard to the amelioration rules. The seller can only *increase* the fixed component, i.e., the initial fixed quote is a *minimum* not a maximum. The buyer can only decrease the fixed bid, i.e., their initial fixed quote is a *maximum* not a minimum. The royalty bid works the same way as in DA1. This institution thus gives the seller the privilege of setting a binding minimum price on the fixed fee, which can be interpreted as a reservation value. Only the royalty can be negotiated downward. Bids cannot "cross" thus when, for example, the fixed bids meet; only the royalty can be negotiated further or vice versa. The same market information is given and the auction starts and ends in the same way as DA1.

In the third mechanism, repeated posted-offer (PO) or "manual Dutch Clock auction," the seller enters a price quote in exactly the same way as in DA1. However, the buyers are limited in bidding space to a simple accept or reject of this offer by clicking on the button labeled "Accept Bid" or "Reject Bid." The information on who or how many have accepted or rejected the offer remains private. If all buyers reject the offer during a bidding round, then the seller can ameliorate the offer by reducing the quote in one or both bidding dimensions just like in DA1. Such amelioration comes at a cost for the buyer. The bid–ask gap is unknown to both buyers and sellers until the quote is accepted by at least one buyer, at which time the binding contract is formed. The auction ends when the *first* buyer accepts the quote in the bidding round or there is a timeout. This mechanism could perhaps therefore be

[18] In the initial sessions, the "sliding ruler" was not implemented and the participants manually entered their expectations, a time-consuming task but with the same result on the calculations. The slider greatly speeded up the trading process and the possibility to quickly check expected profitability of one's own quotes as well as market quotes. "Subjects should make decisions, not calculations." (B. Wilson).

called "manual Dutch Clock in two dimensions" where the "clock ticks" are provided by the seller's reduction in quote in every bidding round. Only the past offers are listed in a ranked order in the public market information.

The secondary market (5) is a standard fixed-price double auction (FP-DA) common in asset experiments where the seller enters a fixed-price quote for the contract in its portfolio, bought in a previous period. The contract terms (fixed plus royalty) agreed upon in the primary market (2) or (3) are not re-negotiated but the contract is transferred "as is." Price quotes can be positive (seller gets money from the buyer) as well as negative (sellers pay money to the buyer). This mechanism allows for already-negotiated terms to be compensated for to a level acceptable to a buyer. The buyer (sellers) can also accept the highest ask (bid), like in DA1. Contracts can be resold in periods 2 and 3. The auction is started as the seller decides to resell (4). The first quote can thus come from either the seller or a buyer.[19] The quotes have to reduce the bid–ask spread. The auction ends when the quotes meet, an "accept" is made or the auctioned is timed out. There is a timeout for the total auction time (90 s) and for the maximum time between quotes (10 s).

Trading occurs over a maximum of 30 rounds, each having three periods, lasting approximately 1–5 min each per round. The experimenter can end the session at any round (in order to limit the session time). Screen displays are updated in real time at any event resulting in change of private or public information.

3.2.3 *Subject Payments, Endowments, and Special "Rules of the Game"*

At the beginning of the experiment, initial roles are assigned to the participants. There is one Inventor (Role 1), one Trader (Role 2), and 6–8 Users (Role 3), with half in "industry A" capable of producing product A, and half in "industry B" capable of producing product B (if an odd number of Users is involved there is one less in industry B).

Each participant is endowed with a capital of experimental money at the beginning of a treatment. The purpose of the capital is to introduce "bankruptcy laws" and "loss aversion." The participants' earnings are decided by accumulated gains (losses) via contract issuing, splitting, and using contracts held in portfolio during each period. At the end of the experiment, participants are paid a weighted sum of accumulated earnings in each role plus an hourly fixed fee (for keeping the capital positive) and a fixed show-up fee (for showing up on time), not counting the experimental money capital endowment. The exchange rate is decided at the end of the experiment (partly due to time constraints on the number of rounds in each

[19] In the later experiments, the market starts by re-listing the contract and either the buyer or the seller can post a first bid–ask. In the earlier experiments, the market started once the seller sent in an ask. The change in procedure was motivated by time constraints for the experiment, and speeded up the resell.

session), and "converged" to 0.1 for Role 1, 1 for Role 2, and 0.5 for Role 3, in order to make all roles approximately equally profitable, which turned out to be a difficult task for most agents.

If the accumulated losses deplete the capital, a one round grace period is given to the participant ("Chap. 11"). If the capital is still negative at the end of the next period, the participant is declared "bankrupt" and cannot participate further in that session, and both the hourly fee and the show-up fee are lost (there is no payment for the session). This seemingly harsh measure was introduced to stop constant overbidding by many participants in the initial sessions, and it had the desired effect. After the rule was introduced, only one person entered "Chapter 11" and no one went bankrupt, although two came very close (compare (Kahneman and Tversky 1979) on prospect theory and "loss aversion."). The importance of the rule in the experiment is that participants who are systematically overbidding "at no cost" destroy the market, which is precisely the dimension we want to study. Therefore, we need to find a solution that gave sufficiently salient incentives for profit-seeking behavior. Simply paying trade gains was not enough (positive enforcement). It was necessary that participants lose something given to them, apparently inducing loss aversion behavior (negative enforcement). The capital endowment solved this problem beautifully. It was therefore not enough to have property rights to trade but also rules that limited the greed of some participants from destroying the market [compare "Thou shalt not steal nor covet your neighbors possessions," and "declaring the fundaments of markets and a warning that distributional jealousy should not be allowed to destroy them" (Smith 2002)].

Participants are informed in the instructions about the linear nature of the contract, the decisions and activities each role can do, the uncertain nature of the sales and blocking profits generated by using the contracts, and the "bankruptcy rules." They can keep the instructions with explanations of the trading screens and experimental flow during the whole session. They are not informed about the distribution of values among participants (which is a linearly increasing function).

In some initial experiments, the secondary market was not used. Invalidated contracts annulled payment obligations (typical practice in real-world). If a contract was invalidated, this information was immediately displayed in the market info box. The number of participants varied between 7 and 10 in all sessions, changing the competitiveness of the demand-side. Each session is reported separately and all data can be related to an individual.

3.3 Overview of Experimental Design, Design Parameters, Market Performance, and the Sequence of the Experiment

I report findings from 24 experimental sessions using the design parameters listed in Table 3.1. A 3×2 design was used for the study of prices and dynamic gains under three primary market institutions and two levels of presumed patent validity (designs 1–6). These were the independent variables.

Table 3.1 Experiments by institutional and legal environment

	Institutional and legal environment		Competition	Contract values			
Design	Institution	Validity (%)	Type A and B users	Mid point of high blocking value: (ABs; ABq, A, B/B+,–)[c]	Values	Experiments	Data
0	DA-1	100	(4+4)	(8;8,4,8/8)	Values1	2p1,5p1	-
	PO	100	(4+4)	(8;8,4,8/8)	Values1	4p1	-
1	DA-1	100/93[b]	(3+3)	(7/8;7/8,3/4,7/8)	Values2	6,7,8,16p2	64
		100/93[b]	(4+4)	(8;8,4,8/8)	Values2	15p2	-
		100/93[b]	(4+4)	(3/4;4,4,0/0)	Values3	9.1, 14p2	27
		100/93[b]	(2+2)	(2;2/3/4,4,0/0)	Values3	13p2	-
		93	(4+3)	(4/9/10;4/9/10,4,0/10)	Values4	17.1	25
		93	(3+3)	(7/12;7/12,3,7/12)	Values5	18.2, 19.1	47
		93	(4+4)	(8/14;8/14,4,8/14)	Values6[a]	20.2	19
		93	(3+3)	(–5/7;–5/7,–5,–9/7)	Values7	22.1	38
2	DA-2	100/93[b]	(4+4)	(3/4;4,4,0/0)	Values3	9.2	14
		100/93[b]	(3+3)	(7/8;7/8,3,7)	Values2	10	10
		100/93[b]	(4+4)	(8;8,4,8/8)	Values2	11.1	24
		93	(4+3)	(4/9/10;4/9/10,4,0/10)	Values4	17.2	11
		93	(3+3)	(7/12;7/12,3,7/12)	Values5	18.1	26
3	PO	93	(3+3)	(7/12;7/12,3,7/12)	Values5	19.2	20
4	DA-1	38	(4+3)	(7/12;7/12,4,7/12)	Values5-38	21.1	34
5	DA-2	38	(4+3)	(7/12;7/12,4,7/12)	Values5-38	21.3	30
6	PO	38	(4+3)	(7/12;7/12,4,7/12)	Values5-38	21.2	12
							401

Cost structure:
Issue cost: Standard contracts 1 and Quality contracts 5 for experiments 2–10 and 2 from experiment 11–21.2
Transaction cost: 0 for experiments 2–17 and 1 for experiment 18–21
Use cost: Variable cost = 60% of sales for experiments 9–17 and 65% for all others
Cost of capital: 30% for Inventors, 5% for Traders, 10% for Users
Insurance cost: 0

Patent renewal cost: 0
Investment: 1

[a] The blocking value range/midpoint is constant in this treatment (constant risk)

[b] validity alternated every 5 periods starting with 100%

[c] The values are given in a range (a, b) to the subjects and a random value is drawn for the realization of the value using a uniform distribution. Other distributions like Poission may be used in future experiments. The midpoint in the range is here shown for the different markets (contract type ABs, ABq, A, B). Quickly common expectations were created for the contract value. In sessions 17–21 there were two alternating blocking values used (separated by the '/') for the AB and B contracts, creating a "two-commodity" market to study robustness in the instiution with respect to pricing on fundamentals

In Exp 1–8 some ABs/q values vary due to rotation among agents with less than 8 on the buyer side, dropping the highest value

(p1, p2) practice/training sessions for team 1 and team 2. Two teams were used to compare. No real differences were found

The induced investing and blocking value ranges in the reported experiments were positive, exploring the linear contract for positive prices.[20] In about half the sessions, there were two B-contract markets, randomly changing between the two values between rounds, designed to test "robustness" of the institutions.[21] The values for the investment and blocking values were given in overlapping ranges, linearly increasing for the buyers to create a competitive bidding environment with only one "high-value" bidder, useful for evaluating allocation performance of institutions.[22]

The Inventor was given a broad range, spanning over the ranges of all buyers, a typical condition for Inventors. The (intermediary) Traders were also given the full range but spanning over each "industry's" ranges. This was based on the presumption that Traders typically have better information than Inventors, being closer to the product market. The Users had the narrowest (most precise) ranges of values being in the market using the technology. Trial sessions were run with valid (100%) contracts to verify institutional performance during the development of the final experimental design and some trial sessions (design 0). A fixed cost of capital, specific to each role type, was used to discount the values over the three periods in each round.[23] Past market experiments suggest that at 3–4 buyers are needed to avoid collusion (some would say 6–8). For the AB contracts, mostly 7–8 were bidding and for the A and B contracts, 3–4. The design thus permitted variation in the competitiveness of demand-side bidding. A homework assignment was given after the final session, asking participants to teach future participants what to think of when trading in different experiments. The purpose with that exercise was to study what participants were *thinking* when trading, as one study what subjects *do* in experiments.

Before the data collection experiments, a prototype test was run testing the *feasibility* of trading a linear contract with uncertain values. The prototype did not include a Trader role but simply tested, successfully, the core of a market mechanism with two-dimensional simultaneous bidding. A special software was developed for this test. After that, a complete parameterization and systemization was made for a first *pilot* set-up. A second software was developed, this time including the Trader role. This tested the dynamic trading including the Trader and showed that the dynamic environment actually worked (the Trader was able to acquire and

---

[20] The linear contract allows "negative" prices in one or two dimensions. Negative fixed price means that the buyer is borrowing funds to explore the rights in the contract for a royalty including the value of the contract + the value of the money (debt). Similarly, a negative royalty means that the seller is paying the buyer for the effort of investing in the contract for a fixed fee including the usage value + the risk in the product market. For role 2, who splits the contract, positive prices in particular mean an injection of capital in the trade (a new use of capital). No voluntary trade was observed with negative prices.

[21] Refer to personal communication with S. Rassenti for introducing the idea of a "robustness" test.

[22] The range was fixed in most experiments ("fixed risk"). In one experiment, 20.2, the range/midpoint value was fixed ("constant risk") to induce more risk for higher values. All realizations of values were done using a uniform distribution across the range.

[23] This variable was thus exogenous in this experiment. Typical cost of capital for the specialiced type of agents were used.

split a contract in a competitive bidding environment). Finally, a *design* and a third, complete, software was developed for the reported experiments. The first experiments using the final design were used as training sessions. Previous experience with complex environments suggests that the experience level and information state of subjects is critical to eliminate the possibility that results are sensitive to these factors.[24] See Appendix A for the recruitment process and a copy of the instructions aimed at qualifying subjects as well as get accustomed to the software. The first sessions thus served as *subject qualification* sessions, perfecting the software for the trading environment, the design mechanisms, and instructions.

Data for value expectations, bids, time series, prices, dynamic gains, and other dynamic system parameters were recorded. A brief visualization of the dynamics in the price adjustment process and of subject's value expectations (risk preferences) is summarized first. The focus is then on the clearing prices and, as a consequence of this impersonal exchange, dynamic gains from trade. Experiments with high validity are reported first (designs 1–3), by institution and market, followed by the experiments with low validity (designs 4–6).

3.3.1 Market Performance and Dynamic Gains Under High Validity

3.3.1.1 Some Observed Market Dynamics in Price Clearing in the Primary Markets

The bidding process is visualized in Figs. 3.2–3.7 for the three market mechanisms: DA1, DA2, and PO. The process clearly suggests that the bidding takes place in both dimensions. Three examples are given. In Fig. 3.2 (ABq/B contract, session 8), the clearing price is paid in fixed fee and royalty that exactly matches the induced blocking and investment values, respectively. The figure also illustrates a "linear" bidding process, i.e., the buyers and seller tend to adjust both dimensions to agree on a price. In Fig. 3.4 (ABq/B contract, session 19.1), the buyers are reluctant to accept a high royalty asked by the seller but in the end bids jump to meet the royalty demand. However, these royalties are at unprofitable levels for investing, indicating that the buyer is intending to use the contract to block. The fixed component clears at just below profitable levels for blocking. In Fig. 3.3 (ABs contract, session 19.1), the opposite happens. The seller eventually drops to clear the price at a lower royalty, but profitable level for investing, presumably intending to Invest. The buyers here appear to signal their intentions by the dimension in which they prefer to increase their bids.

[24] Special thanks to Prof. Vernon Smith for discussion on the importance of pilots and training in complex economic "real-world like" environments prior to data collection. Several pilots were run to verify the economic system design (prototype), the experimental design (pilot) and procedure, software, interface, and instructions (training sessions).

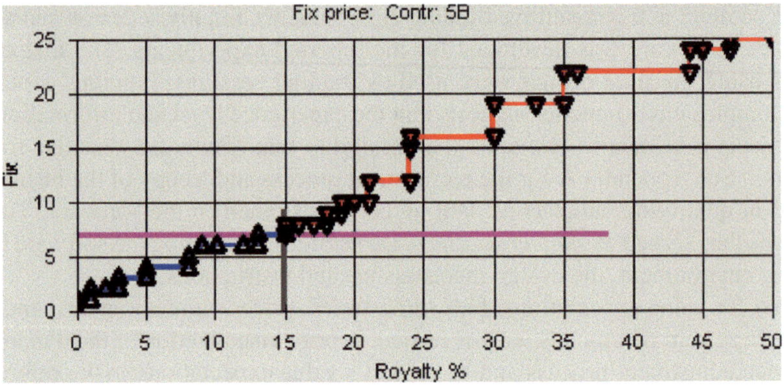

Fig. 3.2 Examples of the bidding dynamics and price dynamics data for DA1 from session 8. *Triangle pointing down* is an ask, a *triangle point up* is a bid, and a *star* is the clearing price. The *horizontal line* is the theoretical fixed price and the *vertical* line is the theoretical royalty if the theoretical fixed fee is paid

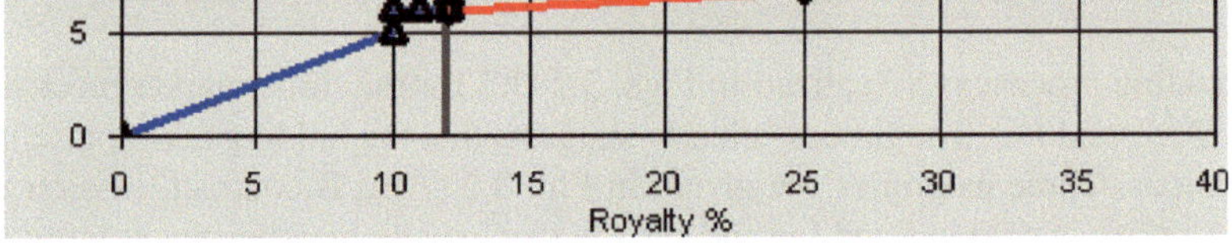

Fig. 3.3 Examples of the bidding dynamics and price dynamics data for DA1 from session 19.1

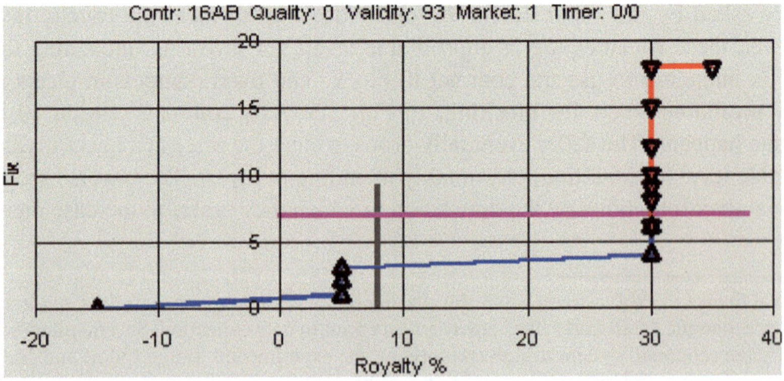

Fig. 3.4 Examples of the bidding dynamics and price dynamics data for DA1 from session 19.1

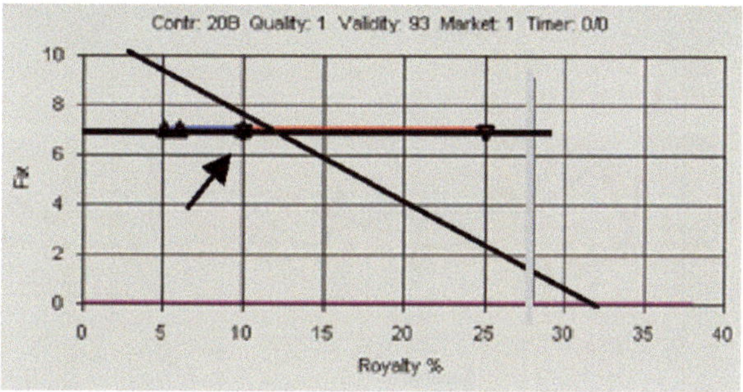

Fig. 3.5 A example of the bidding dynamics and prices dynamics for DA2. The *vertical* and *negative sloped lines* indicate limits to profitable price for blocking and investing use respectively. The arrow and start indicates clearing price

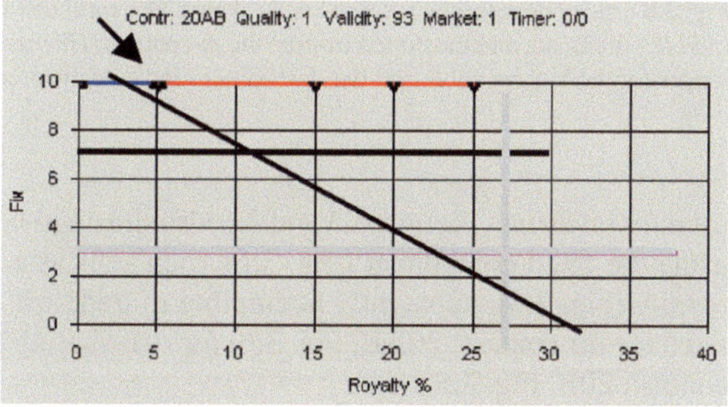

Fig. 3.6 A example of the bidding dynamics and prices dynamics for DA2

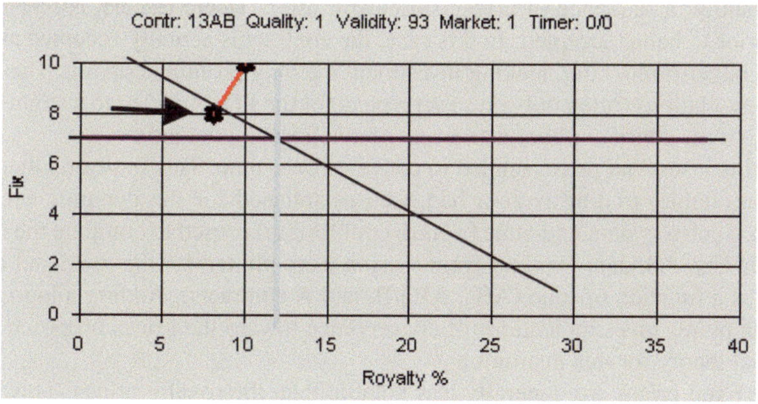

Fig. 3.7 A example of the bidding dynamics and prices dynamics for PO

Similar patterns can be observed in DA2 but with considerably less negotiation. Figure 3.5 (ABq/B contract, session 18.1) shows how the seller sets a reservation value on the fixed fee and then negotiation takes place in the royalty, until the clearing price is profitable for the buyer. Figure 3.6 (ABq/B contract, session 18.1) shows how a high, and clearly unprofitable, reservation value on the fixed fee is traded-off for a lower royalty when profits can still be made from investing without having to accept a negative royalty (never observed). All these contracts were also

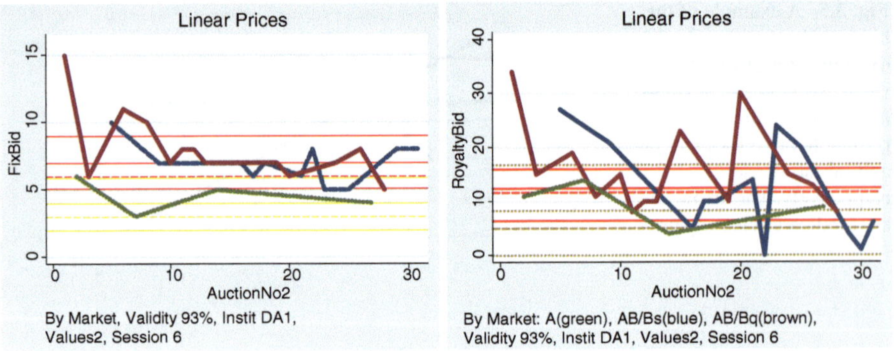

Fig. 3.8 Price convergence for fixed and royalty. The *solid line* (*upper*) is the B/quality, the *dashed* (*middle*) the B/std, and the dotted (*lower*) the A-contract. *Horizontal lines* indicate the induced value ranges for the highest value and the *dashed lines* indicate the predicted competitive outcome

used for investing. Figures 3.5 and 3.6 also illustrate how a *Trader* outbids the Users using the fixed component (Fig. 3.6), then sells at a lower fixed fee, but a higher royalty (Fig. 3.5), apparently attempting to trade-off risk-taking (*accepting* a high fixed fee for contract 20AB, Fig. 3.6) for risk-sharing (*asking* a lower fixed for split contract 20B, Fig. 3.5).

A different pattern is observed for PO. Figure 3.7 (ABq/B contract, session 19.2) which shows a sequence of take-it-or-leave-it offers where the ask is ameliorated once prior to being accepted. In this case, the contract is actually accepted at a loss for the buyer if blocking, making investment the only profitable option. These very short negotiations (often only one ask) typical for the PO, often led to a slight under-pricing of the contract (there was a cost of ameliorating the bid).

During a session, prices tended to converge over time. This indicated that a sustainable (stable) trading pattern had been established for the dynamic economic system. Such a system end-state (a Nash equilibrium) is used to compare the institutions. In Fig. 3.8, data for DA1 from session 6 are plotted for the fixed and royalty prices as a function of time (ABs, ABq/B, and A contracts). Adding a log-trend to the data points gives indications of convergence toward the prices proposed by the informal theory for this institution.

The fixed prices are generally less volatile than the royalty prices. This can be explained in part as a trade-off with the fixed. The average price of the fixed is "epsilon" higher than 6, which equals the theoretical prediction (based on the proposed informal theory). The contracts appear to be priced near predictions. Fixed prices clearly converge to the second-highest blocking value. The royalty price appears to converge on average but is much more volatile.

Contract values were also varied randomly between a high and a low B-contract value (the "robustness test"). Figure 3.9 illustrates the ability of the DA1 institution to price the contracts according to fundamental value (session 18.2). A separation between the B, B+, and A contracts can be seen in the figure, again with a trend to adjust to the theoretical predictions, especially for the fixed price. The B+ contract

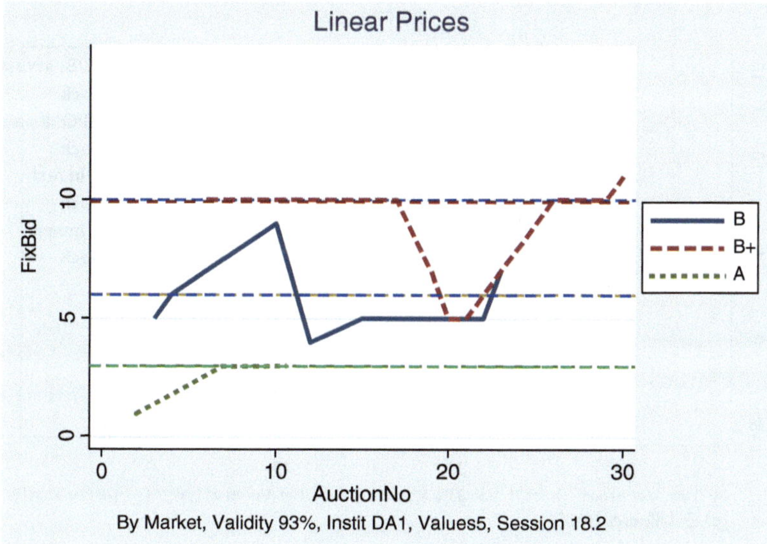

Fig. 3.9 Price convergence for two randomly alternating values for the B-market. The *upper line* is the B+, the *middle* the B, and the *lower* the A-contract. *Dashed lines* indicate predicted competitive outcomes

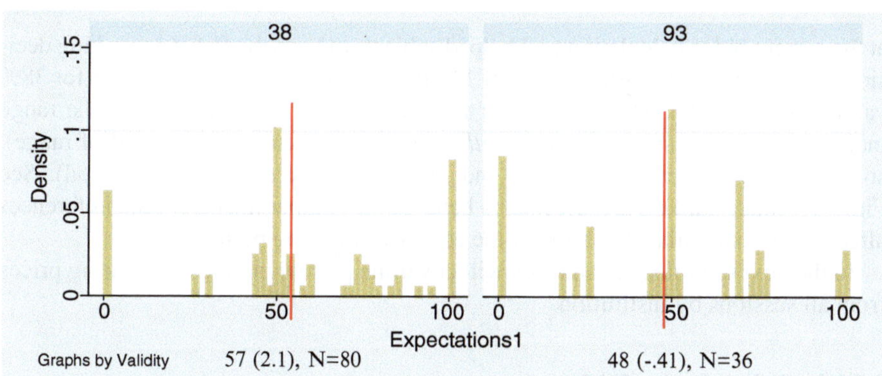

Fig. 3.10 Expectations by validity of the subject winning the contract

royalty prices are higher than expected. These contracts were mostly used to block; thus, no royalty had to be paid, a subject I will come back to later in the analysis.

These examples of initial results appear to support a separation of fixed and royalty prices tied to the blocking and investment value and usage in particular when both dimensions are up for bidding (DA1). Different types of trading patterns were observed for different institutions but these are not reported here except through these examples.

When making their bids and asks, the subjects were asked to decide on the expected outcome of the random draw in the value range (0–100% in the range). These expectations were used by the subjects to calculate expected profits from a contract if acquired

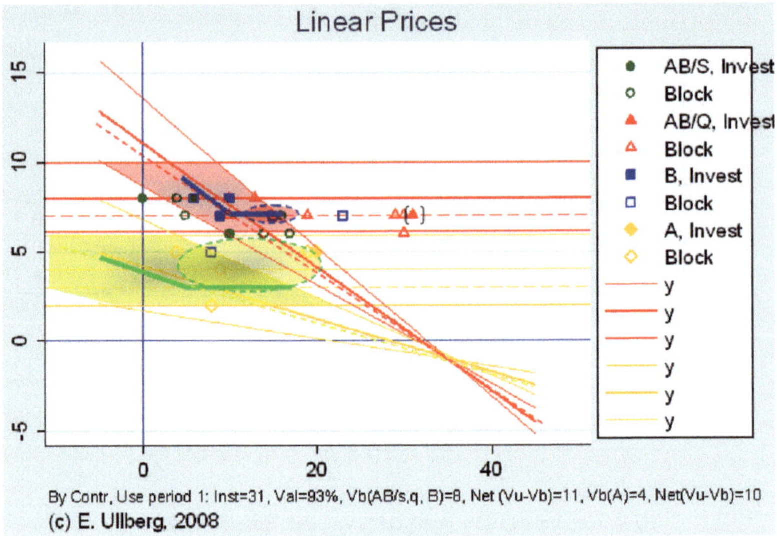

Fig. 3.11 Sessions 6, 7, and 9. Institution DA1. Actual prices for different markets (AB/s,q, A, B) with Max. fixed, royalty price ranges (*solid line*) and *E*(fixed, royalty) prices (*dashed line*). Min. fixed, royalty prices are not indicated for clarity

at the current bid (a calculation tool helped subjects to keep the focus on making decisions, not calculations). An analysis of 116 transactions (sessions 21.1–21.3 for 38% validity and 22.1 for 93% validity) of these expectations of the value in the range indicated that some subjects *consistently* used low expectations (low end of range), some rational expectations (50%), and others high expectations (high end). See Fig. 3.10. Since markets are affected by behavior on the margin, these risk preferences affect prices and initial allocation of the contracts affecting performance.

In the next sections, the analysis will cover the price dynamics of clearing prices from all sessions by institution.

3.3.1.2 Prices Under DA1

Prices for Dual Values

The first experiments with qualified subjects, sessions 6, 7, and 8, were for high-validity patents exchanged using the DA1 institution and are charted in Fig. 3.11. Eight trained subjects participated in each of these experiments: 1 Inventor, 1 Trader, and 6 Users roles, 3 in "industry" with "product mix" of A or B with rotation between roles.

In Fig. 3.11, the linear clearing prices for the contracts traded are plotted with the fixed price on the *y*-axis and the royalty price (expressed as a percentage) on the *x*-axes. The linear prices for each type of market (AB/standard, AB/quality, A and B) are indicated with a different symbol. The symbols that are filled indicate that the contract was used to *Invest* (in period 1) to create a new product (A or B). If they are hollow, the contract was used to *block* (in period 1) to earn a profit from an existing product (A or B). To show profitable bids, solid lines are plotted for the expected profits from blocking and investing based on the upper, middle, and lower end of the respective value ranges (Max. Fix(t) and Max. Royalty(t)).[25] Profitable asks (Min. Fix(t) and Min. Royalty(t)) are calculated in the same way, but for clarity, these values are not drawn in all figures. The lines form a polygon box (shaded) within which prices can be profitable given the uniform random realizations of the ranges. The bold line indicates maximum prices at which contracts are profitable on average. To show expected prices, two dashed lines are used to indicate E(Fix(t)) and E(Royalty(t)|Fix(t)). Where they cross represents the expected clearing price according to the proposed theory of prices. Due to high volatility in prices, mostly in the royalty dimension (this changed some over time in the experiment as subjects became more experienced but consistently remained higher than the fix) and for clarity, an average price is calculated and drawn as a circle/ellipse, ±1 standard deviation from the average.

The average fixed price[26] for the AB/s, AB/q, and B contracts in Experiments 6–8 is 7.0 (28 observations) and prices are narrowly spread around the E(Fix(t)), the midpoint of the second-highest value range (=7). All prices are lower or equal to the midpoint of the highest blocking value range (=8) and higher than the lower end of the highest blocking value range (=6). These results indicate initial support for the proposed theory where the fixed should be equal to the midpoint of the second-highest blocking value. The average royalty price is 13.9% which is higher than what can be explained by the investment value (=11%), but on average just inside the high end of the investment range.

The prices show an interesting dynamic: Most fixed prices are at the second-highest blocking value. The lowest value are at the highest agent's low end, indicating a clear limit of minimum price. The maximum price is at the midpoint of the highest agent. The behavior suggests a dynamic bidding process between the two highest bidders and, on average, an approximate risk-neutral bidding behavior and a clear understanding of minimum profits.

The contracts that are blocked (in period 1) have a higher royalty on average than contracts that are invested in, but there is no apparent difference in the fixed price. Comments like "When I see royalty prices go higher than what is profitable,

[25] Realizing the blocking values will result in horisontal lines (no royalty is paid when blocking). Profits for a user = Vblock(t). t = period 1,...,P. Profits from investment values will result in sloping lines (fixed always has to be paid). The investment value, given as a range of increase in revenues from investing, is converted in the graph to a corresponding maximum royalty as a percentage, minus the fixed price paid, minus cost of goods sold (COGS), and minus an investment each period, related to the new production (and marketing) costs. Profit = Vinvest(t) (1 − COGS − royalty(t)) − fix(t) − Investment per period. Solving Profit invest for royalty(t) gives E(royalty(t)).

[26] In these initial experiments, there was no clear difference between the AB standard, quality, and B contracts. They had typically the same induced value but the strategic use, split for quality, no split for standard, and demand-side competition differed, A and B industry users, or A or B industry users for split contracts, factors that could affect prices. However, a small difference in price, consistent with competitive bidding, is reported at the end of this section for the contracts that were split.

I know that the buyer intends to block" indicate that participants are aware of this bidding behavior. However, prices are not at 50% (the maximum), indicating a restraining behavior in prices when buying to block. Tradability of a contract (prices are profitable for both investing and blocking) may have been one such concern. Being able to price this future value aimed at managing one's own risk in profits from future periods (period 2 and 3) would indicate a pricing strategy prompted by the uncertainty in induced values. Subject comments along the lines "We want to be able to resell the contracts. It's too risky without. One can get stuck with a contract." suggest that the tradability value was in fact part of the pricing strategy, at least among some. Such behavior would be consistent with risk management behavior in liquid markets. Another explanation for restrained royalty prices may be that in the DA1 mechanism buyers can increase the bid *if necessary* to win the contract, but may not always do that initially to keep their intentions private. Interest in re-trading was high during many experiments. Contracts were more or less consistently put up for resale, though these efforts only occasionally resulted in any trades (initial allocation was efficient), and in spite of the fact that this was a time-consuming task, reducing the number of earning rounds in the experiment and reducing earnings for all participants (a behavior consistent with short-term profit-seeking). This prompted the introduction of a small transaction cost, which resulted in more experimental time for primary market trade. Only when unusually "low" or "high" prices were agreed upon (inefficient allocation), then a resell did take place. Positive and negative prices were observed.

But even when investing, the royalty price appears to be slightly "high." This may have an explanation in the design. Since the fixed (and royalty) prices could only be given as whole numbers, a marginally higher royalty (which has a smaller impact on cost than a higher fixed price) might have been used to win the contract, keeping the fixed at the second-highest value. This potential experimental design effect may be reduced in future experiments by using bigger numbers and/or decimal bids.

The average A-contract price is 4.2 (only five observations), which is slightly higher than proposed theory and is close to the midpoint of the highest value range (=4). The average royalty price of 12.0 is also higher than the NPV hypotheses (=10) and just outside the rational expectations "box." Although considerably fewer data points were recorded, it appears that similar pricing preferences can be seen in the prices for A and B contracts. The prices indicate a clear separation in prices with respect to the AB/B and A contract values.

About 30% of the contracts are invested in (70% blocked) and about 80% of the quality contracts (that can be split) are split, leading to an average of 50% of all contracts are split, and thus generate dynamic gains.

Tentative conclusions for DA1 are: the results indicate support for the proposed informal theory for investment use. When the usage is blocking, buyers tend to express their WTP in the fixed (which they have to pay) at a level close to the blocking value but overpricing the royalty (which they do not have to pay in that use). When the usage is investing, buyers tend to express their WTP at royalty levels close to the investment values that can still make a profit, given their WTP for the fix, which again appears to be close to the blocking value. The blocking and investment values thus appear to be *independently* expressed (no linear trend in fixed prices) in the fixed and the royalty dimension to the competitive price level (second-highest values), making the institution incentive compatible (static efficiency). This also suggests that even if an individual buyer intends to Invest, the buyer still has to pay the competitive blocking value.

The number of split contracts indicates the propensity for the Traders to make a profit by first outbidding all industry users and then creating two new "split" contracts with limited rights to each industry A and B, and subsequently sell them one after the other at competitive prices to a User in each industry. The economic value of the initial contract (technology) is thereby "multiplied." The dual values of the A+B markets are potentially realized as a result of trading, increasing the social gains in the system (can be characterized as Pareto gains[27]).

Seeing this market as an integrated market (as in the international trade literature), gains would be generated by the mechanism of impersonal exchange proportional to the propensity to increase the usage of the patented technology to generate market access.

Prices for Random Variation of Contract Values: "Robustness" Test

This separation of the A and B contracts indicates some robustness of the institution to price contracts according to the value. However, since the same trained people returned to the next sessions, common expectations quickly formed *across* sessions. This resulted in a long "hysteresis" effect when the values were changed, lasting most of a session. The "solution" chosen was an extension of the experiment and to use two alternating values for the AB/B contracts where the blocking value was randomly altered between high and low values. This extension, more importantly, provided for a *robustness test* of the institutions at the same time. The question now became whether this institution was also able to "distinguish" in prices, based on fundamental values, between contracts of the same type (AB/B) but with different values (a "two-commodity" market).[28] The results are chartered in Fig. 3.12. In sessions 18.2 and 19.1, using Values5, a high B value (=12) was therefore randomly introduced in the previous value set (B = 7, A = 3/4). I find a clear separation between both the high and the low B-contract values and A-contract values, indicating robustness. The prices generated for the contracts when investing also appear to give additional support to the price hypotheses. However, the prices of the high AB/B-values when blocking appear to follow another pattern. There is a much higher propensity for blocking use, as blocking becomes more profitable (although

[27] Pareto gains is a static concept and for small changes. However, here it is a dynamic context, but as we look at the total system gains, everybody is better off when splitting. This is due to the fact that more market access is created in the process. The incentives are aligned between risk-averse inventor–trader bidding and then trader–innovator bidding (which is not the case between risk-averse inventor–innovator bidding who favors opposite positions on fixed fee and royalty).

[28] This expansion was discussed to some detail with Prof. S. Rassenti, particularly the "robustness" test.

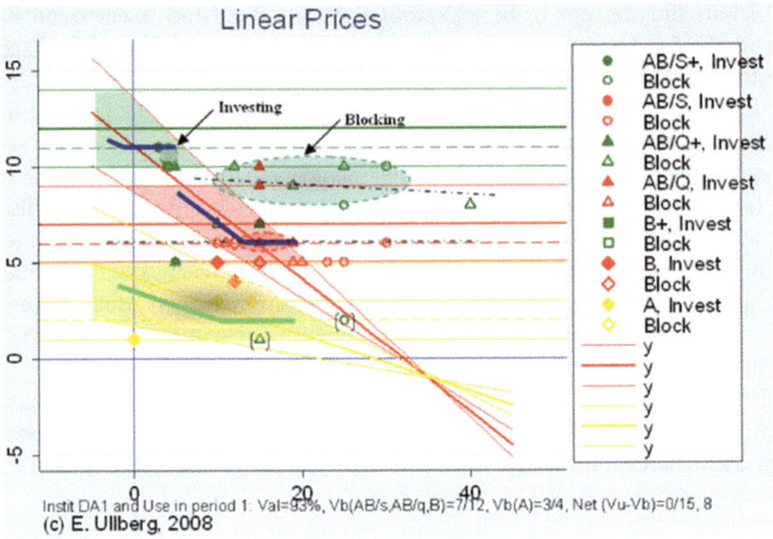

Fig. 3.12 Sessions 18.2 and 19.1. Robustness test of DA1

the same profits could be achieved by investing). The royalty prices are considerably higher for the blocking use than the investing values (as in previous sessions). This decision of a "safe," but potentially lower, profit appears to be consistent with risk-averse behavior. If one uses the prices from participants with much experience, >5 sessions, the "investing" prices are within the "rational box." The blocking values are however still "undervalued" but close to the lower blocking value range. In a similar way as the previous experiments, A-contracts on average seem to be slightly overvalued.

By accepting a high royalty, only high-value "blocking" buyers are able to participate, reducing the competition and possibly colluding with the remaining high blocking value buyers, reducing the fixed fee further. The tradability option in periods 2 and 3 appears not to be sufficient to restrain overbidding in the royalty price when an "easy" profit can come from blocking. Yet another possible factor for these prices is that at least some participants "average" the values of the contracts, i.e., form common expectations for both high- and low-value contracts. This "rule of thumb" pricing was observed in comments by *some* participants, i.e., "I always start at (10.15%) when bidding," whereas others appeared to try to get the "fundamental value" prices. There appear to be competing bidding strategies at play simultaneously. Also, only blocking is profitable except for very low royalties and some lower value "blocker" may bid up the royalty and thus force high-value Users to bid for blocking use (royalties are overlapping).

I conclude that similar results are found in these experiments as in the first. It seems that the hypothesis of the fixed price is strengthened if the use is investing (no obvious linear trend in prices). The uncertainty in the values is not enough to restrain agents from "destroying" the tradability value by bidding a high royalty, at least not when the values are given for all three periods at once. If given only one

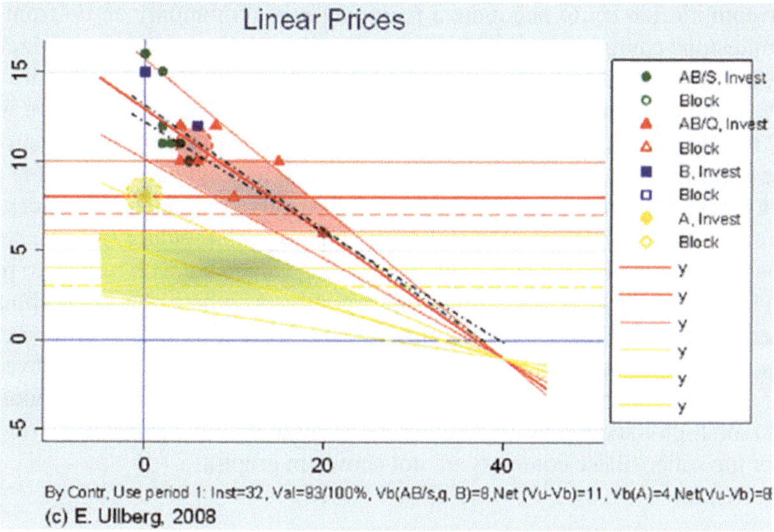

Fig. 3.13 Sessions 10, 11.1. Institution DA2. The royalty prices lay close to the investment values, given what is paid in fixed. The *upper dash-dot line* includes all prices and the lower only nonsplit contract prices

period at time, perhaps that would change the expectations of future earnings and this behavior (which may be closer to some naturally occurring environments). About 25% of the contracts are used for investment and about 28% of all contracts and 40% of quality contracts are split and generate dynamic gains.

3.3.1.3 Prices Under DA2

Prices for Dual Values

Once we had learned something about prices using DA1, the second institution was tested to compare the prices using a reservation value on the fixed component, reducing the bidding space to the royalty dimension but having the same public information. In experiments 10 and 11.1, Values2 were used with institution DA2 and chartered in Fig. 3.13. The contracts are significantly higher in the fixed price, consistent with risk-averse behavior by the sellers. However, the royalty prices appear to be fully compensated to a royalty that makes the contract profitable for investing. All contracts above the middle blocking value are used for investing, consistent with the fact that blocking is not enough to make a profit on average. The higher fixed fee changes the *incentives* for the buyers, investing more in the contracts, generating royalty and fixed income for the Inventors. In real-world contracts, the Inventors often try to negotiate a fixed payment (commonly an up-front lump sum, milestone payment, or some minimum payment scheme) to "incentivize" buyers to make efforts to invest in the contract, overcoming the moral hazard problems.[29] In the experiment, the proposed interpretation is that the buyers are willing to pay the blocking *value* in the fixed fee and trade some of the investment profits to get the contract, but only if they have to. In contrast to previous sessions, the prices show a clear linear trend, making the fixed and royalty components dependent. The slope of the trend is almost identical to the highest value bidder (not the second-highest), indicating that the buyers are overpaying somewhat with respect to predictions. However, these prices include split contracts, where the price should be higher. Taking out these contracts may put the regression closer to the second-highest value. Two other observations can be made: (1) the propensity to Invest was higher than DA1 with about 60%, but slightly less split contracts traded, about 20%; and (2) too high fixed reservation values also lead to several "no trades," which is a loss for the seller (these contracts are not shown in graph).

Prices for Random Variation of Values: "Robustness" Test

As was done using DA1, a session with random AB/B contract values were also run for DA2 to study robustness of this institution. The results from session 18.1 are chartered in Fig. 3.14. The considerably higher fixed prices are consistent with Experiments 10 and 11.1, but the data clearly appears more noisy and with less apparent separation between the high and low B-contracts. Many prices are also profitable. In part, this may be explained by experience. If outliers are excluded on the basis of lack of experience and the investing and blocking values are distinguished from each other, there seems to be a separation of prices between high- and low-value B-contracts. The average price for the high-value B-contract is within the "rational box" for the investing use. The blocking use is lower than the minimum value, just as for DA1. The same reasons may apply here as for DA1. Again, only one A-contract was traded and the price is clearly higher than the blocking value and more in line with the B contracts. The data set is, however, small (seven observations/value).

The conclusion from these two experiments seems to reinforce the tendency under DA2 for higher fixed prices. The royalty prices are then (linearly) compensated to create a profitable contract for investment. The fixed-price hypothesis appears not to hold in this one-dimensional bidding. An auxiliary hypotheses then becomes: When a reservation value can be set on the fixed fee, the prices follow a linear pattern, compensating for a fixed fee > blocking value by reducing the royalty. The gains from trade as expressed by split contracts are low in DA2.

---

[29] Interview with one of Sweden's most successful inventors, Karl Derman, confirmed that this incentive was critical and without the fixed payment, priority may be given to other projects in the decision-making process.

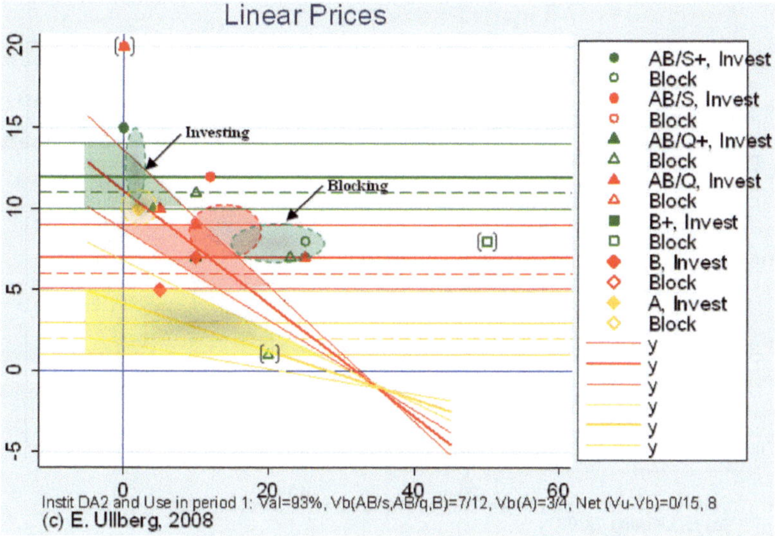

Fig. 3.14 Session 18.1. Robustness test of DA2 with high-validity contracts

3.3.1.4 Prices Under PO

Prices for Random Variation of Values: "Robustness" Test

Finally, the PO institution was tested using random AB/B contract values and is chartered in Fig. 3.15 (session 19.2). The PO institution appears to be considerably less able to distinguish the differences between the high and low values. In fact, in this session, common expectations appear to have formed for both contract values at the lowest value level. This may alternatively be explained by the riskiness of exploring high values, since there is limited information on WTP (only accept/reject of sellers WTA). Prices appear to be slightly lower than what are predicted by the hypotheses, but are expressed in both the fixed fee and the royalty near the competitive price. Also none of the contracts were split, which means that there were no dynamic gains from trade. An interesting incentive appears to be that prices are discovered in a way in which both the fixed and royalty are changed, showing a weak positive linear trend in the prices [the slope of a regression of the prices is positive (not statistically significant result)]. This line intersects near the rational expectations price.

A tentative conclusion seems to be that prices are lower than in the other institutions, which shifts profit from the sellers to the buyers. Since this institution is the one closest to today's *personal* exchange, the result suggests that high-value technology may be undervalued due to the lack of competitive demand-side bidding.

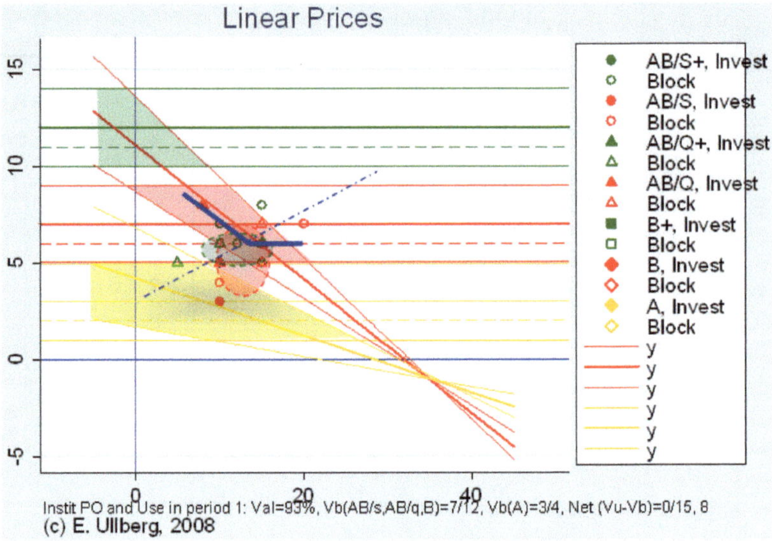

Fig. 3.15 Session 19.2. Robustness test of PO with high-validity contracts

The supply side is unable to fully discover the WTP of the demand side without demand-side bidding. It is interesting to note that this institution – the most similar to today's trade – results in 70% blocking (only 30% of the contracts were used to Invest), which is a common critique from Inventors when they sell their patents to Innovators.

3.3.1.5 Prices for Split Contracts Under DA1

A last look at the prices for high-validity contracts, now differentiating between contracts that are split and not split is chartered in Fig. 3.16 (sessions DA1: 6, 7, 8 and DA2: 10, 11.1). Split contracts give slightly higher prices in fixed than nonsplit contracts for DA1 and higher fixed with lower royalty for DA2, indicating that Traders outbid Role 3 using the fixed component. Outbidding the highest User's values then puts the competitive price at the highest value (the upper fixed line), which can be seen in the figure. The Trader turned out to be a rather difficult role for most participants. The fixed price is close to 8 (7.8) for the split contract, the highest Role 3 value, whereas it is close to 7 (6.8) for nonsplit contracts, the second-highest Role 3 value. This indicates that the Inventors benefit in the fixed price from the increased value created by the Trader splitting the contract. The dynamic profit incentives are actually aligned between the specialized agents to maximize the use of the technology: Inventors with a tendency to be risk-averse (as seen in the DA2 where a higher fixed free is a behavior consistent with risk aversion) prefer fixed, but Users prefer royalty for the same reason (they do not

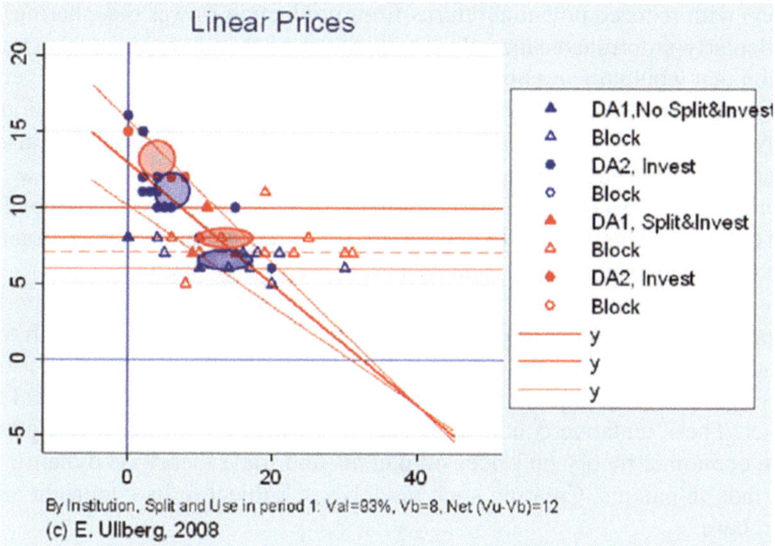

Fig. 3.16 Session DA1: 6, 7, 8 and DA2: 10, 11.1. Split prices appear to follow the same value logic, where Traders (Role 2) on average bid close to the highest User's (Role 3) values. DA1, DA2, PO for high-validity split contracts

want to expose themselves to a high fixed, especially not if investing). The Trader can take on a higher fixed fee by potentially "multiplying" the fixed fee from the two split contracts. The Trader can then accept a lower fixed fee for a higher royalty from the Users, something that is in the interest of the Users. This mechanism thus provides the system with higher dynamic gains based on trade and risk-sharing in a larger consumer market. The role of a Trader therefore appears critical in increasing dynamic gains from trade in patents.

3.3.1.6 Price Comparisons Between Institutions

I have now examined results from the three institutions with respect to prices and gains. Is there a consistent pattern of prices that can be seen? The incentives created by different rules give very different dynamic price patterns. DA1, the only institution with two-dimensional bidding, appears to price both the fixed fee and royalty, in support of the proposed theory (especially for contracts invested in). DA1 also has the highest number of split contracts, i.e., the highest dynamic gains with respect to the use of technology (for Invest and Block). DA2, with one-dimensional negotiation, has clear incentives to generate high fixed fee prices, but at the cost of lower royalty and fewer split contracts (less dynamic gains). This means that the sellers have to pay for the lower risk they want (by asking a high fixed fee increasing risk transfer) with reduced potential returns from production (lower risk-sharing). This is particularly unfortunate since almost all contracts then have to be invested in, a situation that would *favor* a high royalty agreement for the sellers. The buyers also appear to overpay in royalty. PO, with no fixed fee or royalty bid information from the buyers, has the lowest volatility in prices (they are "set" by seller), but also fails to price high-value contracts according to fundamentals, making it the lease robust institution. It is also a rather costly venture for the seller to discover prices by failing to sell contracts. This appears to push prices down to a common expectation across contracts at the lower contract value level (the price risk is high).

The tentative conclusion from the high-validity experiments gives experimental support for the proposition of division of the fixed and the royalty with respect to blocking and investment value: one needs two-dimensional institutions to express the two-dimensional value independently in the two-dimensional linear contract. These tentative conclusions may be fruitful for further investigation to inform economic theory on prices on patents and trade theory on dynamic gains from trade in patents. Creating such models is a subject in its own right and not treated here.

3.3.2 Experiments with Low Validity

3.3.2.1 Prices for DA1

Prices for Random Variation of Values: "Robustness" Test

The next group of experiments *repeated* the study of prices for different institutions but with "low" validity at 38%, which introduces substantial uncertainty in the contract value, especially for investing use. In the environment with random high/low AB/B values, the "robustness test" was used. See Fig. 3.17.

The results paint a picture similar to that of high validity, particularly with regard to fixed prices. The prices are separated clearly with the lower B-contract around the blocking value and the royalty below the lower end of the value range. Also, almost all contracts of this type were invested in. The higher B-contracts are lower than the lower end of the blocking value and priced higher than the highest royalty value. All contracts of this type were used to block. The A-contracts are priced higher than the blocking value but about the middle of the investment value.

The profits from investment are calculated in an industrial approach, with three production batches during one period, generating inventory to sell. If the contract is invalidated, the remaining inventory is lost, generating a loss proportional to up to one-third of the revenues for that period (similar to receiving a court "injunction"). It therefore appears that this risk of loss of inventory and profit is reflected in lower royalties, i.e., less risk-sharing, as buyers try to compensate for a potential loss. The patent system can therefore be seen as a "risk management institution" for the trading partners, whose actions are important for the risk-*sharing* in the contracts.

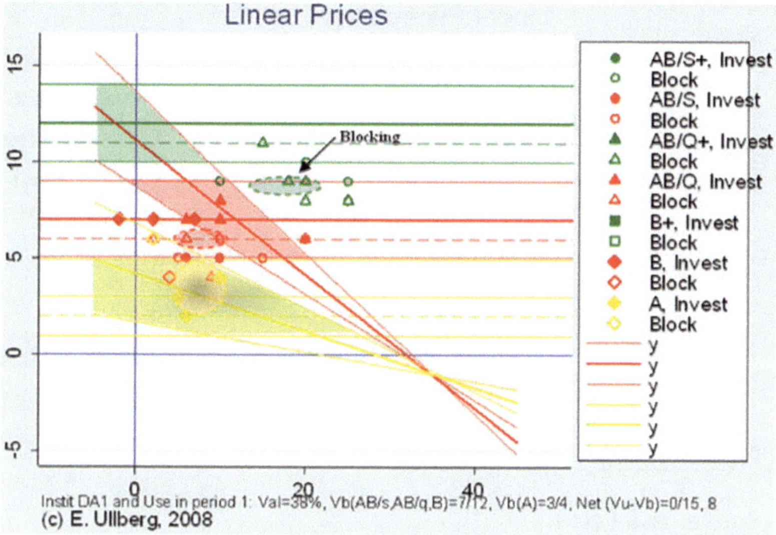

Fig. 3.17 Session 21.1. Prices for DA1 under low validity

A tentative conclusion is that the hypothesis on pricing of the fixed is further supported. However, for the high blocking values and low investment values, subjects appear not to discover the most profitable use, but again "settle" for a safer blocking strategy. The lower validity appears to lower the royalty price, consistent with compensating for this risk of loss from production. The prices of the contract appear again to be related to the usage. The propensity to Invest is around 20%, which is considerably lower than for high validity. The propensity to split contracts is 25%, which is comparable to high validity.

3.3.2.2 Prices for DA2

Prices for Random Variation of Values: "Robustness" Test

The DA2 institution prices the values quite differently under low validity. Almost all contracts are used to block. See Fig. 3.18. The B/B+ contract prices are indistinguishable and common expectations appear to have formed near the higher end of the lower B contract value. This can be explained by the fact that the increased risk in investing makes buyers unwilling to trade-off a high fixed for a royalty, "forcing" Users to block. In fact, the only time Users invest is at a royalty rate of 0, i.e., low risk with respect to royalty payments. It may also be explained by the sellers not asking high fixed prices (they did not learn how to trade in this environment), giving no incentive to trade-off between fixed fee and royalty. An additional study of nontraded contracts may reveal if this is the case or if buyers

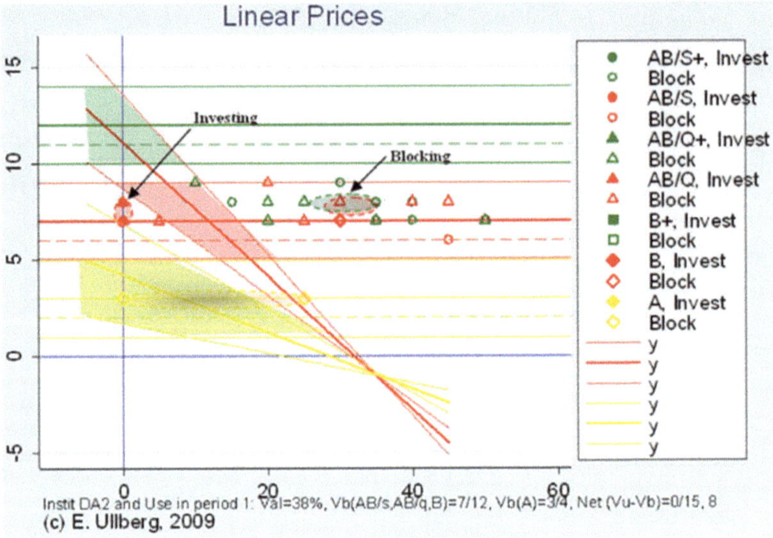

Fig. 3.18 Session 22.3. Prices for DA2 under low validity

simply do not accept a high fixed fee. The (only two) A-contracts are, however, priced closer to fundamental value.

The two contracts traded for investment clearly have a lower royalty than under high validity (the only contracts recorded with 0% royalty). This result indicates that the adjustment in the price from variations in validity is done in the *royalty*. This is probably motivated by the risk of losing the investment value in the low validity treatments. This is thus a result consistent with DA1. This session was run last (high experienced participants) and in the middle of Experiment 2 (with the same institution). These results may therefore only be indicative and a re-run could possibly give a different result. Here the stable end-state is clearly blocking. The propensity to Invest is around 10%, which is considerably lower than for high validity. The propensity to split contracts is around 20%, which is comparable to high validity.

3.3.2.3 Prices for PO

Prices for Random Variation of Values: "Robustness" Test

The prices for PO are not possible to distinguish between high and low AB/B contract values, a result similar to high-validity contracts. See Fig. 3.19. The difference appears to be in a lower royalty, like the other institutions. The propensity to Invest is around 33%, which is considerably lower than for high validity. The propensity to split contracts is around 8%, which is also low compared to high validity.

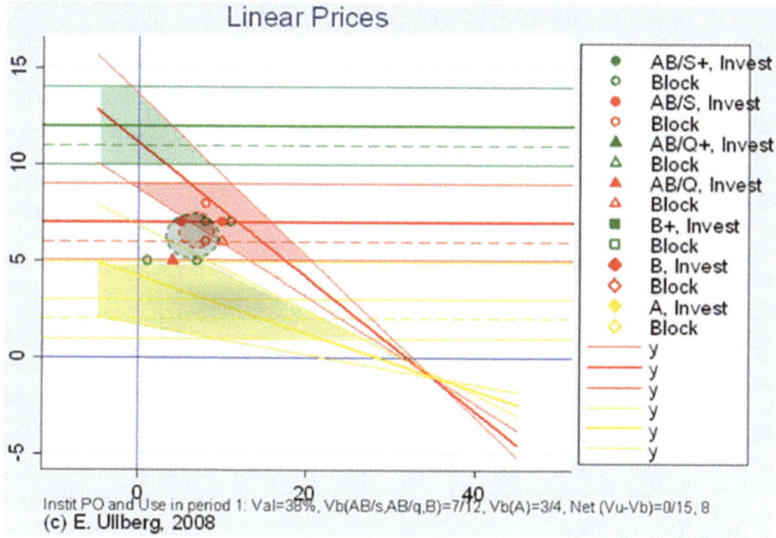

Fig. 3.19 Session 22.2. Prices for PO under low validity

When comparing the average prices between the three institutions for the lower validity, there is a consistent trend toward lower royalty prices (which one would expect). The royalty prices are about 15–30% lower with about 60% lower validity (93%→38%). A tentative conclusion is that institutions seem to have the same price patterns (same incentives) and the change in the legal environment reduces the WTP, a royalty apparently to compensate for the risk of loss of production (inventory). This compensation takes its expression in the royalty only, not the fixed fee.

3.3.2.4 Prices for Split Contracts

The prices paid for the split contracts for the three institutions under validity 38% are chartered in Fig. 3.20. The split contract prices are slightly higher, as in the case of 93%, here shown for DA1. The incentives for trade appear to be unchanged with respect to validity.

3.3.2.5 Comparison Between Institutions

Comparing the price dynamics for the institution under low validity indicates a more common behavior among the institutions than under high validity. The fixed prices are generally independent of the royalty. This can be explained by the higher propensity for blocking (possibly due to uncertain value to Invest); therefore, it is basically the blocking value that is priced. As noted, the DA2 session was run after

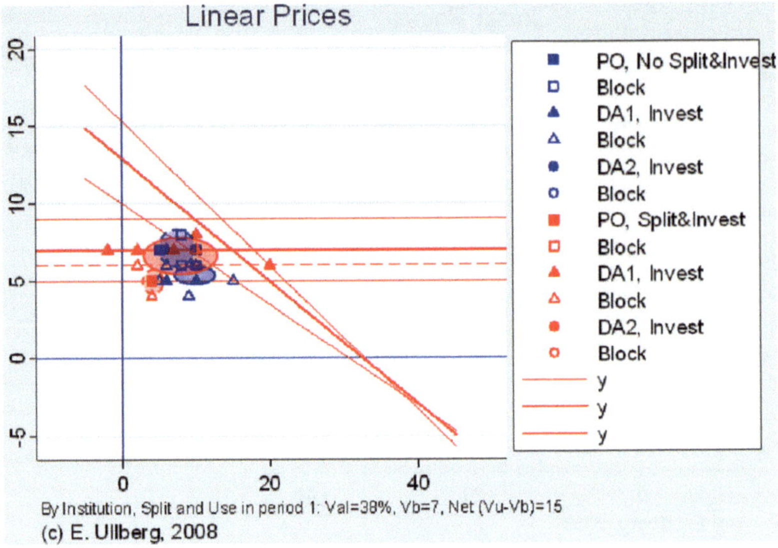

Fig. 3.20 Session 22.1-2. Split prices under low validity. DA1, DA2, PO for low-validity split contracts

the others and may result in a more pronounced trend if re-run. The indication is that with low validity, mostly the blocking value is priced. This is clearly an indication of a different end-state of the system than for high-validity patents, with possible implications for patent policy.

3.3.3 Secondary Market Reallocation

The secondary market was used in periods 2 and 3 to reallocate contracts, for a fixed price, bought in the primary market, for a linear price (fixed fee plus royalty). Only a low frequency of actual trades was observed, indicating that the initial allocation was perceived to be efficient. See Fig. 3.21 where allocation of contracts is chartered with respect to highest value holder and under different competition (six or eight bidders).

Ten percentage of primary market transactions (in period 1) resulted in a reallocation (in period 2 or 3). Of these, only 6% had a price different from 0, i.e., the transfer was accompanied with a price to be paid. Only 2% were transacted under low validity (Fig. 3.22).

Even though actual transactions were few, the propensity to choose to resell was high, eager to maximize profits at the cost of trading fewer contracts during the experimental session. Basically, every round contract holders first listed the contract to resell. Since the purpose of the experiment was to study the primary market, a

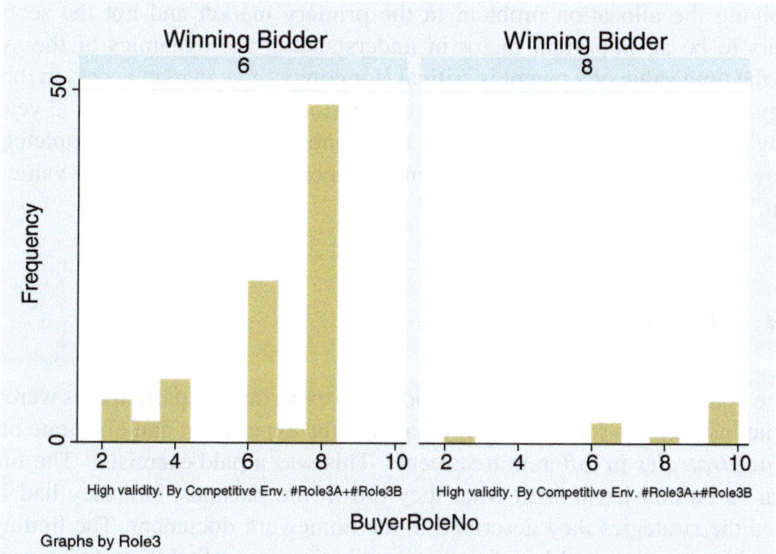

Fig. 3.21 Sessions 18.2–22.1. Winning bidder for contracts under different competition. Role 3A = 1–3–5–7 and Role 3B = 2–4–6–8–10

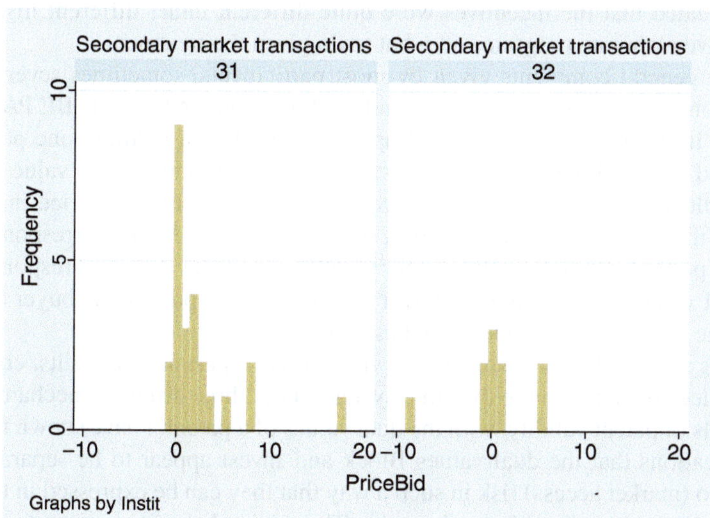

Fig. 3.22 Sessions 6–21.3. Secondary market prices by primary markets DA1 (=31) and DA2 (=32). No re-trades were made for PO

transaction cost of 1 was introduced which reduced the propensity to list contract to those that were perceived to be initially underpriced. This minimal transaction cost was sufficient restrain overuse of this function and was necessary to make the experiment and the system more efficient.

Solving the allocation problem in the primary market and not the secondary appears to be an important factor in understanding the dynamics of the system, since the time value of a patent is critical (for competitive market access in the technology market) and limited. A typical "trial period" of a technology is 1–3 years and getting it wrong the first time may cut the economic life in half or completely take it away.[30] A well-functioning primary market therefore adds to the time value of the patent, a dynamic effect of the system.

3.3.4 Homework

For the homework assignment at the end of the experiments, participants were asked to write instructions for future participants in the experiment and elaborate on their *bidding strategies* in different treatments. This was a paid exercise.[31] The findings appear to be consistent with what they did in the sessions, i.e., they had implemented the strategies they described in the homework document. The findings are reported as statements. Main factors contributing to profitable performance were noted as: the competition (how strategic the other participants were in their bidding) and the different treatment conditions (institution, validity). Comments like "It is important to adapt your strategy to different conditions in order to maximize profits," indicated that the incentives were quite different under different institutions. Some gave elaborate programs on what to do given the treatment.

Some general comments given by most participants, sometimes several times, was: "Don't be greedy," "Slow and steady will get you ready," and "BE PATIENT," possibly indicating experience build-up over time. When bidding, one participant expressed the maximum price to bid as: "As a rule of thumb, pick 1 value less than the middle value of the blocking range as your highest bidding price in order to increase the chances of making a profit of more than 50%." This corresponds to the smallest possible profit to make greater than 0 on average. It also corresponds to the midpoint value of the competing buyer just below, thus putting the buyer at risk of losing the contract to the second-highest value bidder.

In this section, I have given an overview of the experimental results, comparing linear prices obtained in the dynamic system under three different mechanisms and two levels of patent validity, with the dual values of a patent. I have shown that there are indications that the dual values Block and Invest appear to be separated with respect to (market access) risk in such a way that they can be expressed in the linear contract dimensions fixed fee and royalty. The pricing behavior, contract usage, and dynamic trade gains also vary among the institutions.

---

[30] Average maintenace time of a patent is about 7 years, varying from 0 (chip industry) to 30 years (pharma with patent extention time). For most technologies, initial efficient allocation would therefore be critical, making the institutional design a critical factor affecting the price.

[31] This idea was suggested by Prof. S. Rassenti.

3.4 Linear Prices, System Gains, and Potential Capital Allocation: Hypotheses and Empirical Results

In formulating some hypotheses on the linear prices, I will distinguish between the predicted prices based on the proposed informal theory (nomothetic experiment) and the heuristic study of the messages and information, both under different institutional arrangements and legal validity. The first approach studies under which particular institutional and legal settings convergence to the predicted outcomes is more likely. The second approach aims at discovering differences in price outcomes between different institutions and legal environments. Capital allocation is a simple calculation based on the dynamic gains as an indication of systemic risk between treatments.

3.4.1 Linear Prices Compared with Proposed Theory (Static Analysis)

I begin with the question of predicted static "equilibrium" prices versus observed prices. The theoretical proposition predicts that the willingness to pay a fixed price is equal to the second-highest blocking value plus epsilon and the willingness to pay a royalty price is equal to the second-highest investment value, net the fixed price already paid, plus epsilon.[32] The question of how accurate the predicted prices are with respect to the observed prices is examined using a *t*-test of the difference between the predicted values and the observed values

$$\text{Fix} - \text{Fix}^{*} = 0 + u, \quad \text{Fix}^{*} = \text{Vb}_{-1}, \tag{1a}$$

$$\text{Royalty} - \text{Royalty}^{*} = 0 + u, \quad \text{Royalty}^{*} = \left[(1 - \text{COGS}) - \left(\text{Vb}_{-1} + \text{Investment}\right)/\text{Vi}_{-1}\right], \tag{1b}$$

where fixed and royalty (expressed in percentage) are the clearing prices and predictions (* superscript), Vb_{-1} is the midpoint of the second-highest blocking value range, Vi_{-1} is the midpoint second-highest investment value range, COGS are cost of goods sold (1−COGS is then the "gross margin"), investment is the periodic investment if

[32] "The second-highest plus epsilon," since the induced values are discrete and the highest bidder only needs to outbid the second-highest bidder with epsilon. This is true if the same agent has both the highest blocking and investment values. Otherwise, prices will be between the second-highest and the highest values (not studied here).

investing in the contract (development, marketing), and u is the random error term.[33] Predictions are accurate in both price components if the calculated differences in (1a) and (1b) are both 0. The hypotheses under rational expectations then imply that (1a) and (1b) cannot be rejected. Table 3.2 lists the means tests of (1a) and (1b) with t-values by institution, legal environment, and markets.

These results indicate a strong support for the predictions of the fixed price for the B1/B2 markets (contracts with investment and blocking values) and DA1, PO institutions. These results hold for both levels of validity. When blocking, the royalty is overpaid, but not drastically. The tendency to overpay is consistent with an intention to block (no royalty paid), but the tendency to do so with some moderation is consistent with risk-neutral behavior (to be able to resell the contract if it does not turn out to be profitable). For DA2, the tendency is to overpay in fixed but also to consistently trade-off the higher fixed with a lower royalty.

In particular, the fixed prices are within 0.5–10% of the predictions across these markets. A test of fixed = 0 and royalty = 0 is firmly rejected. The "robustness test" with random alteration of B+ and B contracts show that DA1 slightly underpays fixed and overpays royalty with high propensity to block; DA2 slightly underpays fixed and overpays royalty even more (also with high propensity to block); and PO clearly underpays fix, overpays royalty, and only blocks. The "responsiveness" to different validity indicates that DA1 prices are adjusted downward in the royalty (as the uncertainty in investing increases), DA2 overprices the fixed, and PO does not adjust any prices (overpaying the royalty).

---

[33] Simplified predicted values can be calculated from the equations

$$\text{NPVblocking} = \left[\text{Fix} - \text{Vb}_{-1}\right] \times \text{Sum}\left(1/(1+d)^t\right), \tag{2a}$$

$$\text{NPVinvesting} = \left[\text{Vi}_{-1} \times (1 - \text{COGS} - \text{Royalty}) - (\text{Fix} + \text{Investment})\right] \times \text{Sum}\left(1/(1+d)^t\right), \tag{2b}$$

where NPVblock/invest are the net present values from blocking and investing over the three periods. If the NPV from using the contract should be positive or zero (full truth revelation), solving these for fixed we obtain

$$\text{Fix} = \text{Vb}_{-1}, \tag{3a}$$

$$\text{Fix} = \left[\text{Vi}_{-1} \times (1 - \text{COGS}) - \text{Investment}\right] - \text{Vi}_{-1} \times \text{Royalty}, \tag{3b}$$

This solution assumes that Vi and Vb are constant over all three periods. In most cases, Vb was constant but Vi was increasing. A solution taking discount rates into account was used for the actual experimental comparisons. In the demand-side competition (a Bertrand type of competition), prices would be defined at the intersection of these two lines. Substituting (3a) in (3b) we then get the following two means tests

$$\text{Fix} - \text{Vb}_{-1} = 0 + u, \tag{4a}$$

$$\text{Royalty} - \left[(1 - \text{COGS}) - (\text{Vb}_{-1} + \text{Investment})/\text{Vi}_{-1}\right] = 0 + u, \tag{4b}$$

where u is the random error term.

Table 3.2 *T*-test of means for different contract markets over institutions and validity

Legal environment				Institutional environment								
				DA1			DA2			PO		
Econ.env.		Vb-1		*f–f**	*r–r**	*N*	*f–f**	*r–r**	*N*	*f–f**	*r–r**	*N*
				(t-test)	(t-test)		(t-test)	(t-test)		(t-test)	(t-test)	
93/100	A	3	Inv	**1.9** ***	**−7.9** ***	23	**6.9** ***	**−4.9**	7			
				(5.56)	(−4.8)		(5.35)	(−1.5)				
			Blk	**1.1** ***	**2.9**	18	**5.6** ***	**−2**	5			
				(2.47)	(0.87)		(3.73)	(0.31)				
			Tot	**1.6** ***	**−3.1**	41	**6.3** ***	**−3.7**	12			
				(5.58)	(−1.65)		(6.68)	(−1.17)				
	B1	6	Inv	**0.35**	**−0.4**	17	**4.6** ***	**−2.3**	11	**0.5**	**1.4**	4
				(1.03)	−0.37		(3.29)	(−0.77)		(−0.48)	(0.55)	
			Blk	**−0.03**	**8.6** ***	21	**8**	**−9.2** ***	2	**−0.25**	**5.7** ***	4
				(−0.13)	(5.41)		(1.33)	(−3.7)		(−0.33)	(2.32)	
			Tot	**0.1**	**5.3** ***	46	**4.9** ***	**−3.4**	13	**−0.4**	**3.5**	8
				(0.53)	(4.17)		(3.63)	(−1.28)		(−0.63)	(1.93)	
	B2	7	Inv	**0.57**	**−0.8**	7	**4.3** ***	**−9.3** ***	15			
				(1.18)	(−0.56)		(6.64)	(−6.8)				
			Blk	**0.4**	**12.5** ***	22	**−1**	**2.9**	1			
				(1.48)	(4.25)							
			Tot	**0.45**	**9.3** ***	29	**4** ***	**−8.6** ***	16			
				(1.9)	(3.74)		(5.75)	(−5.75)				
	B+	10	Inv	**−1.4**	**9.5** ***	5	**2.5**	**7.4** ***	2			None
				(−1.25)	(4.53)		(1)	(4.4)				
			Blk	**−2.3** **	**20.9** ***	13	**−3** ***	**29** ***	7	**−4.3** ***	**18.5** ***	11
				(−2.64)	(5.9)		(−2.64)	(6.2)		(−11.14)	(11.5)	
			Tot	**−2** ***	**17.7** ***	18	**−1.8**	**24.3** ***	9	**−4.3** ***	**18.5** ***	11
				(−2.97)	(6.2)		(−1.42)	(5.02)		(−11.14)	(11.5)	

(continued)

Table 3.2 (continued)

Legal environment	Econ.env.	Vb-1		DA1 $f-f^*$	DA1 $r-r^*$	DA1 N	DA2 $f-f^*$	DA2 $r-r^*$	DA2 N	PO $f-f^*$	PO $r-r^*$	PO N
				(t-test)	(t-test)		(t-test)	(t-test)		(t-test)	(t-test)	
	B++	12	Inv			None						
			Blk	**−2.4** ***	**36.0** ***	5						
				(−6)	(5.5)							
			Tot	**−2.4** ***	**36.0** ***	5						
				(−6)	(5.5)							
38	A	3	Inv			None			None			
			Blk			None	**0**	**4.2**	2			
								(0.33)				
			Tot			None	**0**	**4.2**	2			
								(0.33)				
	B1	6	Inv	**0.57**	**−4.7** ***	7	**1.7** ***	**−8.6** ***	3	**0.33**	**−2.22**	3
				(1.33)	(−2.24)		(5)	(−5.5)		(0.5)	(−0.72)	
			Blk	**−0.5**	**0.3**	10	**1.5** ***	**22.4** ***	12	**0.67**	**0.11**	3
				(−1.63)	(0.12)		(5.7)	(5.04)		(1)	(0.125)	
			Tot	**−0.06**	**−1.8**	17	**1.5** ***	**16.2** ***	15	**0.5**	**−1.1**	6
				(−0.21)	(−1)		(7.1)	(3.34)		(1.17)	(−0.69)	
	B+	10	Inv			None			None			None
			Blk	**−1.5** ***	**21.8** ***	13	**−2.3** ***	**36** ***	15	**−3.8** ***	**16.4** ***	5
				(4.16)	(8.34)		(−12.5)	(10.02)		(−7.75)	(9.4)	
			Tot	**−1.5** ***	**21.8** ***	13	**−2.3** ***	**36** ***	15	**−3.8** ***	**16.4** ***	5
				(4.16)	(8.34)		(−12.5)	(10.02)		(−7.75)	(9.4)	

	Blk	170	105	42	23
	Tot	**274**	**164**	**80**	**30**
			36%	48%	23%

f = fix bid, f^* = predicted fix bid
r = royalty bid, r^* = predicted royalty bid
B markets = ABs, ABq, and B contracts, i.e. split and non-split are included in the statistic.
t-test significance in rejecting the null hypothesis ($f–f^*=0$; $r–r^*=0$): *** = 99%, ** = 97.5%, * = 95%

Table 3.3 T-test of means for non-split and split contracts for institution DA1

				Institutional environment: DA1								
Legal environment				Non-split				Split				
Econ.env.		Vb-1	Vb	*f*–**Vb1** (t-test)	*f*–**Vb** (t-test)		*N*	*f*–**Vb1** (t-test)		*f*–**Vb** (t-test)		*N*
93/100	B	6/7	7/8	**−0.02** (0.12)	**−1.08** (−6.55)	***	**48**	**0.7** (2.34)	*	**−0.44** (−1.44)		**27**
	B1	6	7	**0.03** (0.14)	**−1.06** (−4.64)	***	**32**	**0.28** (0.65)		**−1** (−2.31)	*	**14**
	B2	7	8	**−0.125** (−0.62)	**−1.125** (−5.58)	***	**16**	**1.15** (2.96)	***	**0.15** (0.39)		**13**
38	B1	6	7	**−0.56** (−1.89)	**−1.56** (−5.29)	***	**9**	**0.5** (1.18)		**−0.5** (−1.18)		**8**

B market refers to B1 & B2 markets
f = fix price paid
Vb1 = Second highest blocking value, Vb = Highest blocking value
There is a tendency to trade off royalty for higher split fix prices; not statistically significant
t-test significance in rejecting the null hypothesis (f–f^* = 0; r–r^* = 0): *** = 99%, ** = 97.5%, * = 95%

This behavior thus appears consistent with prospect theory (Kahneman and Tversky 1979): A decision under uncertainty would give priority to preservation of the value (fix/insurance). The investment value can be seen as being priced as a strategic option (royalty/gambling). If the decisions under uncertainty are made in the observed way, then this is also in support of the proposed informal theory to price the fixed component as an insurance. These results indicate support for the hypotheses of the blocking value expressed in the fixed price, if the prices can be expressed independently, and the royalty, the difference between the investment value and what is paid in fixed, is expressed as a percentage.

Auction theory and experimental studies suggest that prices converge over time close to equilibrium no matter what the institution (with one-dimensional prices). Here, they do not appear to do that, although the linear prices appear to share a solution (they cross near the same price, which is close to the predicted price for high validity). The differences in linear prices between institutions appear to be stable. Furthermore, since patents are limited in time, the speed of convergence for different technologies matter. An inability to distinguish between contract values (the robustness) translates to the inability to discern between high- and low-valued technology in the "real-world." If one would rank the institutions based on the *t*-test statistics on mean prices: DA1 > PO > DA2 in *accuracy*, DA1 > DA2 > PO in *robustness*, and DA1 > DA2 > PO in *responsiveness* to changes in validity.

To test for price predictions on split contracts, the same *t*-test of means is made, comparing split and nonsplit contracts, using DA1 data (highest propensity to split). Table 3.3 re-confirms the fixed price hypotheses, with price differences of less than 0.5% (!) from predicted fixed prices for high-validity contracts, when the split

Table 3.4 Instiutional comparison with respect to dynamic efficiency
*** INITIAL CALCULATION ***

Market		Institution					
Legal and Instit Env.		DA1	*N*	DA2	*N*	PO	*N*
93/100	**No-split**	26%	59	**37%**	30	19%	14
	Split	**40%**	13	31%	6	0%	0
	Total	28%	72	**36%**	36	19%	14
38	**No-split**	**19%**	16	16%	13	18%	7
	Split	16%	2	**29%**	3	0%	0
	Total	**19%**	18	18%	16	18%	7

Bold numbers indicate highest values of dynamic efficiency (realized consumer surplus/optimal endstate surplus)
Demand side bidding doubles the dynamic efficiency from 19% (PO) to 40% (DA1)
Weak patents reduce performance back do around 19% across institutions

contracts are taken out of the average. This result suggests *identification* of the fixed fee with the blocking value.

3.4.2 Comparison Between Institutions (Dynamic Efficiency)

The trading system is a dynamic microeconomic system. To calculate a "dynamic" efficiency measure, the optimal end-state (as in dynamic programming) is used in the denominator. The numerator is the actual path taken through the system in the three periods in each round used, together with expected values at each decision point (the average value in the induced value range). The realized expected value is then divided by the optimal path's expected value. An initial calculation is made using data from the three institutions and two validity treatments. See Table 3.4.

The distribution of efficiency measures differ clearly between split and nonsplit contracts and the three institutions. The ranking is DA1 > DA2 >> PO for split contracts and DA2 > DA1 > PO for nonsplit contracts. The dynamic gains are thus clearly realized more for DA1. An explanation for this is that the values are more independently expressed in the DA1 prices; therefore, the price risk is lower, making Role 2 (Trader) simpler and increasing the propensity both to split and to create quality contracts. Further analysis will reveal more information on efficiency and rankings.

The results are unsatisfactory for institutional efficiency. One would like to see values of 95%+. This result suggests that efficiency in impersonal exchange in patents is a difficult task to achieve for subjects. Additional institutional design work is needed to match comparable real-world institutions (like stock markets). The results therefore provide an important clue as to why there are no organized markets yet. It is simply a difficult task. The recent attempts toward this goal have been made by intermediaries, who bring buyers and sellers together. This is the

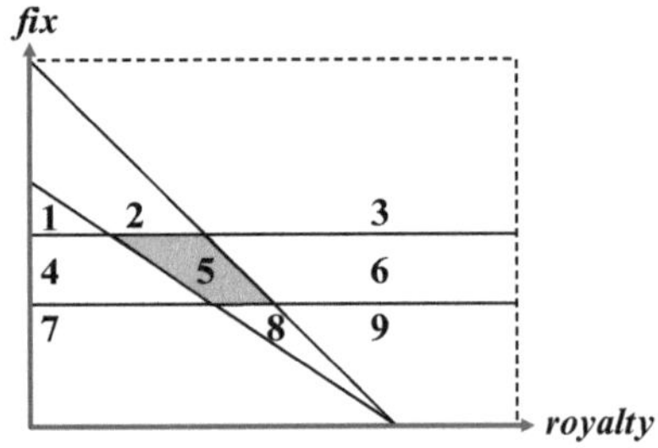

Fig. 3.23 Numbering of the "price zones" or "boxes" used in the statistical treatments. Number 5 indicates the predicted value

same pattern as during the time of the royal privileges, and particularly after the 1836 patent reform cited earlier. However, the stock market did not start overnight; private exchange, especially during IPOs is still very much used. Wall Street started due to a rain that forced people under the tree in the park where they met to trade. Here, they overheard each other's bids and asks and decided *everyone was better off* if these bids were written on a board that everyone could see. We are still waiting for that rainy day in the park when it comes to patents and buyers. Sellers may still be uncertain if they will do better with public prices.

3.4.3 Comparison Between Institutions (Heuristic Analysis)

The second question involves a heuristic study of the incentives (rules), messages, and information the institutional arrangements have on prices. This statistical study is designed to put the *future* into the linear prices. The complexity here is that there are two dimensions in the price. The bidding space is divided into 9 two-dimensional "price zones" defined by the minimum and maximum predicted values of the blocking and investment value ranges. This division turns the price data into a one-dimensional distribution of prices.[34] The first three areas are divided for the blocking value: below, in, and above the blocking value range. The same is done for the investment value (expressed in royalty). Thus, we have three boxes of royalty values for each fixed value, forming nine boxes (compare the price data figures in section 5). They are then numbered from left to right, starting with the top boxes. This creates the numberings of the bidding space shown in Fig. 3.23.

The frequency of the number of prices in each box is calculated and becomes the one-dimensional dependent variable, more easily suitable for statistical analysis. Number (5) is thus the predicted value, "at the money"; (4) indicates underpaid in royalty, "in the money"; and (6) "out of the money." Number (1) is overpayment in fixed an underpayment in royalty; (2) is "at the money" in royalty and overpaying fixed; and (3) is "out of" the money in both prices. Number (7) is underpayment and

[34] Ruling out a treatement based on a system of equations, an extensive search in methods to examine two-dimensional problems did not result in any "typical" treatment, in fact no examples were found.

Table 3.5 Institutional comparison with respect to prices

Pricing: B1 market		Price zone											
Legal and Instit Env.		1	2	3	4	5	6	7	8	9	*N*	χ^2 statistic	
93/100	DA1			1	5	22	11	2		1	42		
												0.007 ***	
	DA2		2	2	2	2	1				9		**0.449**
												0.223	
	PO				1	4	1	2			8		
38	DA1				5	2	2	1			10		
												0.067	
	DA2				2	2	10				14		**0.096**
												0.007 ***	
	PO				1	5					6		
Robustness: B1/B+ market		**Price zone**											
Legal and Instit Env.		**1**	**2**	**3**	**4**	**5**	**6**	**7**	**8**	**9**	***N***	**χ^2 statistic**	
93/100	DA1			2	8	34	27	5	4	4	84		
												0.001 ***	
	DA2		6	5	5	6	2			4	28		**0.002** ***
												0.006 ***	
	PO				1	4	1	4	8	1	19		
38	DA1				5	2	4	3	1	8	23		
												0.161	
	DA2				2	2	10		1	14	29		**0.011** ***
												0.000 ***	
	PO				1	5		2	3		11		

X^2 statistic rejecting the null hypothesis (institutions are similar): ***= 99%, ** = 97.5%, * = 95%

"in the money" in both prices; (8) is "at" the money in royalty and underpayment in fixed; and, finally, (9) is underpaying fixed and "out" of the money in royalty. There are two tests performed.

1. *First*, a simple χ^2 is calculated pair wise for different institutions by validity and market. This "similarity hypotheses" means that the null cannot be rejected if there is no difference in pricing incentives between the institutions. Results are shown in Table 3.5.

   The numbers in the table indicate the χ^2 probability. DA1 and DA2 show different pricing patterns for high validity, both in the B market and under robustness tests. The proposed explanation is that a seller, not wanting to lose

Table 3.6 Validity: probability change of prices for predicted validity

Price zone	38%	→	**75%**	←	93%
4	0.36		**0.27**		0.22
		−25%		+25%	
5	0.29		**0.44**		0.51
		+50%		−15%	

(loss aversion again) a profitable blocking value, sets it on the high end, but not too high, to negotiate a trade-off with the buyers for a royalty that is still profitable if investing. Under low validity, this difference disappears, as investing becomes more uncertain and buyers appear less willing to take the risk to make the trade-off. DA1 and PO differ in robustness, as PO appears not be able to distinguish between values under these conditions. In a "stable" technology environment (B-market only/high validity), DA1 and PO are similar; however, PO also underprices contracts (7). Since PO most closely resembles existing bilateral negotiations, high-value technology may be underpriced in today's personal markets as a result. DA2 and PO are also different except for stable technology environments.

2. *Second*, a multinomial logistic regression is run with the prize zones as dependent variables and six explanatory variables. These are: type of institution, level of validity, type of contract, contract quality, predicted fixed price, and predicted royalty price, given what is (actually) paid in fixed. The *future* is thus in the price (prize box) through controlled treatment and environmental variables, and the *present* is controlled for by using the fixed and royalty given the (actual) fixed price negotiated by the subject. The *past*, such as subject experience and past round prices, was not found to yield significant coefficients. Using the regression coefficients to *predict* probabilities of the price zones by institutions and validity confirms the general dynamics that can be observed in the figures, means, and χ^2 analysis. However, additional information can be drawn from the regressions in predictions; for example, the model may be suitable for predicting key aspects of patent and patent market policy or patent *trade* policy. Such policy issues are altering the presumed patent validity or the predictions, which gives an *absolute* measure of the changes in risk-bearing (fix) and risk-sharing (royalty) expressed in changes in the probability of a price zone. Running the predictions for a validity of 75% (about twice the low validity, 38%) yields an increase in *royalty* prices for underpaid contracts (zone 4) to predicted price (zone 5) with 50% probability (from 29% to 44% probability). This translates in Inventors getting paid more (in royalty) by strengthening the patent system in a weak area (e.g., EU). Similarly, this translates to a reduction of prices (zone 5) in the high-validity area (e.g., USA) of −15% (from 51% to 44%). See Table 3.6.

Thus, *harmonizing* the patent system between a strong and a weak patent region (patent laws probably end up somewhere in between) yields gains for the weaker region and losses for the stronger region.[35] However, strengthening the patent system in the weak area while *maintaining* the superior strength in the strong area (no harmonization) may well allow the areas to *trade* in technology, increasing returns on technology as "patent assets" get *similar* validity. The "technology price gap" in terms of predicted prices (zone 5) goes from 22% (.51–.29) to 15% (.51–.44), a 30% reduction. The market access risk of returns from technology per se is reduced for the stronger area's firms. Thus, having *competing* patent systems may strengthen *trade* in technology and lift the weaker area, while also lowering risk in market access for the stronger area.[36]

3.4.4 Gains from Trade (Dynamic Analysis)

Do we observe dynamic gains from trade, expressed by a propensity to split contracts? In formulating some hypotheses on systemic gains, I propose that a richer bidding language would yield more reliable prices as the price dimensions can be expressed more freely, which would lower the risk for Traders when selling split contracts. The institutions with the richer language would then yield the higher gains. The number of *dimensions* of demand-side bidding is 2, 1, 0 for the institutions DA1, DA2, and PO. The propensity for: (1) splitting; (2) investing; and (3) investing in quality contracts are calculated for each institution. A ranking of the institutions is then made.

1. Using data from experiments studying the B1, B2, and B+ markets ("robustness test" data), DA1 yields a propensity to split *quality* contracts of *31%* ($N=39$, t-test for split=0 yields 4.1***), and DA2 yields *12%* ($N=16$, $t=1.46$) and PO *0%* (no split contracts). The t-test gives an indication of the propensity. I conclude that DA1 > DA2 > PO in dynamic gains measured in split quality contracts.
2. If we now include the outcome of the usage decision (Invest or Block) and use the propensity to Invest, the *socially preferable* systemic gains defined as propensity to split and Invest are: *10%* for DA1 (with propensity to Invest=35%, $N=55$, $t=5.33$ for Invest=0), *9%* for DA2 (Invest=75%, $N=20$, $t=7.54$), and *0%* for PO ($N=7$, Invest=29%, $t=1.54$). This indicates that the propensity to Invest, which has been determined to be generally higher for DA2 (see Table 3.2), compensates the lower gains from trade and approach DA1. The ranking, however, remains DA1 > DA2 > PO, even when the social preferable outcome is taken into account. For low validity, the propensity to split is similar; however, including the propensity to Invest yields *7%* for DA1, *1%* for DA2, and *0%* for PO, which tells the same story – DA1 > DA2 > PO. Since DA2 appears not to sustain the trade-offs between fixed and royalty under low validity (presumably due to risk

[35] This argument has similarities in *predictions* to Krugman's "Technology Gap" model for international trade, where a case for "technological protectionism" is argued, but different in *analysis*.

[36] This proposition is in line with the historic development and use of patent systems to attract technology and technology development.

of loss of contract), only the blocking option remains (again, see Table 3.2). I conclude that two-dimensional bidding allows the prices to adjust in such a way that investment remains an option even for low validity, making a proposed ranking of DA1 > DA2 > PO for dynamic, socially preferable gains from trade.

3. The decision to Invest in a quality contract by the Inventor is a voluntary decision in the experiment. In part, it signals a willingness to cooperate with Traders. This factor is thus part of the dynamic gains. The propensity to Invest in quality contracts is 51% for DA1 (N=39), 50% for DA2 (N=16), and 37% for PO (N=7). Including this decision in the gains we get *5.3%* for DA1, *4.7%* for DA2, and *0%* for PO; thus, the ranking is the same: DA1 > DA2 > PO. The agents are thus able to coordinate investment in quality contracts, splitting, and using to create dynamic gains in a *voluntary* environment.

3.4.4.1 Organized Markets Increase Dynamic Gains

It is also worth noting that PO, which most resembles today's bilateral exchange of patents, appears not to have the same incentives to sustain dynamic gains. This would suggest that organized markets, at least operating under the principles implemented in this experiment, would increase the socially preferable use of technology in the economy. With two-dimensional bidding (DA1), an increase in the use of technology of *30%* (through split contracts) is observed and propensity to Invest in technology including the split contracts increases with *10%* in the system. Since the linear contract is the most common contract, and Inventors in some cases do achieve a fixed component, the results indicate that a substantially more socially favorable outcome could be realized through two-dimensional bidding.

3.4.5 Capital Allocation (Calculation)

In the experimental system, the agents are endowed with a discount rate for their values as a measure of the riskiness of the activity. The discount rate can be seen as a cost of capital for the type of agents in the economic system. A calculation of the average cost of capital in use is used as a measure of systemic risk (nondiversifiable risk in a portfolio sense). However, in the contract trade, risk in the technology is shared in a systemic way, including market access risk for new and existing products and services and not just financial risk. This reduces the uncertainty in returns from the technology. The capital allocation then directly reflects the gains.

The cost of capital in the system, calculated as the average of Inventor (30%), Trader (5%), and User (10%) capital, is compared between the institutions based on their propensity to split contracts (share the risk in the product market). A calculation gives: 18% for DA1 (30% split contracts), 19% for DA2 (12% split), and 20% for PO (no splits) both for high and low validity. As an indication of possible dynamic gains from trade using PO as a benchmark for today's personal exchange, the DA1 institution would reduce the systemic financial risk by *9%* and the DA2 by *4%*.

3.5 Summary and Conclusions: Prices and Dynamic Gains from Trade

The conclusions are summarized under headings:

3.5.1 Linear Prices and Informal Price Theory

1. For the DA1 institution and the markets where contracts have both investment and blocking values (B-markets), the results are in strong support of the informal price theory for the *fixed fee* of the linear contract, i.e., the fixed fee should be *equal* to the blocking value of a patent. This result is particularly strong when the split contracts are taken out of the mean prices and then differ less than 0.5% from the predicted value ($t=0.12$ for rejecting the null that fixed=Vb_{-1}). I conclude that what is priced in the fixed fee appears to be the blocking value of a patent. For the low value markets (A-markets), there appears to be a different pricing pattern for DA1 with what appears as a "floor" for the fixed fee close to the cost of creating a contract. The royalty is traded-off to compensate for the higher fixed fee, similarly to DA2.

   For the DA2 institution and high validity, the fixed fee is consistently *higher* than the blocking value, but never below it, and the NPV of investing is within positive territory, indicating that a trade-off with royalty takes place to compensate for the higher fixed fee. This trade-off disappears for low validity.

   I conclude that the fixed fee includes the blocking value, but for high-validity patents it also includes part of the investment value that is traded of a lower royalty (when the investment option is more certain).

   For PO, the fixed fee is *lower* than predicted on average, but not significantly lower. This institution is closest to the personal and bilateral exchange common in today's patent markets, indicating that current market mechanisms without demand-side bidding would tend to under price the patent.
2. A fixed fee appears to always be paid (fixed=0 is rejected for all markets), leaving the "residual" value to be expressed in the royalty. For the DA1 institution and contracts with dual values, the results support the predictions of the royalty for *investment* use. The royalty is significantly higher than predictions for *blocking* use, making them unprofitable for investing. However, the royalty is still somewhat restrained, in spite of the fact that there is no payment of royalty necessary for blocking. An explanation is that a buyer may want to leave open the option to resell the contract, consistent with portfolio risk management through the secondary market, or might not want to reveal his intentions when bidding, consistent with strategic bidding behavior. Thus, there appears to be "two" prices, depending on how the contract is intended to be used. This result is contradictory to common criteria for a pricing system where the price should be the same independent of its use. In this experiment, the values were given for the

three periods up-front, reducing the uncertainty in the expected value ranges over time. If values were given only one period at a time (more like real market conditions), the investment option might be more important and the royalty "blocking price" might be closer to the "investment price."

I conclude that when contracts are invested in, the royalty paid is consistent with the predicted price, and when contracts are blocked, they exceed profitable investment options at the examined level of uncertainty over time. For the DA2, the royalty is depressed when investing as a result of trading off for a higher than average fixed, sometimes insufficient though, resulting in losses. Since almost all contracts were invested in under DA2, there are no reliable results for blocking. For PO, royalty prices weakly support the predicted prices for investing. For blocking, the royalty prices are higher. Thus, blocking yields unprofitable prices for investing in most cases.

3. The pricing pattern expressing a preference for fixed fee over royalty appears to be consistent with prospect theory (Kahneman and Tversky 1979), where an "insurance" (blocking strategy) is preferred to "gambling" (investment strategy). This result also gives support for the informal theory on linear prices proposed where the blocking value appears to be priced with the fixed fee as an insurance contract (risk transfer) and the investment value with the royalty (risk-sharing). The price for taking over short-term market access risk (blocking) and long-term market access risk (investment option) are thus negotiated in the price of the contract. In a similar analysis using Arrow-Debreu securities (Arrow 1962), the two risks are separated into a fixed transfer price and an option price on "states of nature," that could be derivated on location, time, etc. The result is that there is no risk in the transfer price and all risks are spread over option markets for *each* state of nature. In the experiment using linear contracts, these risks are thus *in* the linear price.
4. Traders win contracts by outbidding the users, primarily using the fixed fee. This is consistent with the proposed theory that Users are unwilling to pay more than the blocking value and lose a strategic option of blocking if the fixed fee is too high. The Traders recover the higher fixed through the sale of the two split contracts. This suggests that Inventors (who create quality contracts that can be split) and Traders (who split) are able to coordinate their activities through the market mechanism to enable more efficient risk-bearing and risk-sharing in the dynamic microeconomic system. I conjecture that both the incentives of an impersonal exchange and a reliable price mechanism with public prices yield Pareto gains.
5. Boundary conditions of moral hazard result in re-pricing of the contracts to profitable levels for use in *one* dimension, investing or blocking. This suggests that voluntary trade occurs only for positive values of investing and blocking.
6. Secondary market trades only occur when primary market prices are too low, which was a rare event. This result shows that the primary markets tested are perceived as efficient in allocation of contracts in the dynamic system. No secondary market trades occurred in the PO treatment however, possibly resulting from the of fact that prices were lower than for DA1 and DA2, i.e. more profitable for buyers.

3.5.2 Institutional Differences in Price Dynamics

1. Stable trading patterns emerge for different institutions and validities, indicating different end-states of the dynamic microeconomic system (Nash equilibriums). This nonconvergence to a common price suggests that demand-side bidding rules matter in pricing the linear contract and that the richer the language/information (measured by the number of bidding dimensions, 0, 1, or 2), the more accurately the linear contracts on patents are priced under rational expectations compared with the informal theory.

   I conjecture that the most accurate prices can be observed when the demand-side can express the fixed and royalty dimensions *independently*. This suggests that two-dimensional simultaneous price negotiation is preferred in pricing the two-dimensional contract. Any limitation of bidding space therefore leads to trade-offs given institutional rules and validity with less accurate prices as a result.
2. Different legal environments (validity) mostly affect the willingness to share risk. A high validity encourages risk-sharing, i.e., use of technology by investing in innovations. There is a shift in prices toward lower royalty prices for low validity. The fixed fee appears not to be affected as measured by price areas. This result may have implications for patent policy with respect to trade. A patent *trade* policy would include a focus on the price dynamics for fixed (transfer) and royalty (sharing), given changes in the legal environment. In particular, the claims structure, which decides which rights are granted, may be the attention of such policy so that the rights become easily divisible (field of use, geography, and time) and tradable (contractual law link and validity in the divisible dimensions).
3. The treatments with randomly varying values in the B-markets ("robustness test") yield significant differences in the prices between institutions DA1, DA2, and PO as measured by price zones. The single B-market, where the technology environment can be interpreted as being "stable" (like marginal inventions), yields less of a difference between DA1 and PO, but differences remain between DA1 and DA2. Since PO is closest to today's mostly bilateral exchange markets in patents, i.e., personal nonstandard licensing agreements with no public prices, differences between institutions suggest that fixed and royalty prices for inventions will be higher for both investment and blocking values under an impersonal exchange mechanism with demand-side bidding, resulting in a lowering of the risk in inventive activities.

3.5.3 Predictions of Prices

1. A multilogical regression analysis of price areas allows for prediction of prices under different validities. A strengthening of a patent system from 38% to 75% presumed validity under the DA1 institution would yield a 50% increase of the royalty prices in the rational price area for the Inventors. Such a policy prediction suggests that the risk in inventive activities goes down. The same prediction, now compared with a patent system of 93% validity, yields a reduction in prices in the rational price area by −15%.

This suggests that a one-sided policy change to strengthen a weaker system may be preferable to a policy of harmonization, motivated by gains from trade. Harmonization may result in trade losses for the initially stronger system. Strengthening the patent system may promote trade in technology between the developing world and the developed world (similar to products and service). Key implication for the patent system may lie in the claims structure.

3.5.4 Gains from Trade

1. Risk in inventing is lowered through the higher prices from impersonal exchange and high validity and split contracts with higher fixed yield (to outbid Users). Therefore, this is a shift in risk-bearing and risk-sharing toward the User and ultimately the consumer, which changes the incentives of capital allocation in the economy to benefit inventive activities. I conclude that such a potential increase in inventive activity could be beneficial to economic growth based on development of technical ideas.
2. The dynamic gains from trade, expressed as propensity to: Invest in quality contracts, split contracts, and Invest in contracts, yield DA1 > DA2 > PO. For split contracts, the values are 30% (DA1), 12% (DA2), and 0% (PO). The highest dynamic gains may be a result of a "reliable" institutional mechanism, reducing the price risk in reselling the split contracts. Such considerations may also be a motivational factor for the Inventor to make the extra investment in a splittable quality contract. The dynamic gains including the socially preferable outcome of investing in the contracts yield 10% (DA1), 9% (DA2), and 0% (PO).
3. Gains from trade between different agents with different cost of capital would lead to a more efficient allocation of capital in the economic system. The propensity to split contracts reduces the systemic risk given the multiple users of technology and the lower cost of capital for Traders (5%) and Users (10%) compared to Inventors (30%). A calculation comparing gains with PO (no gains) yields a reduction of –4% for DA2 and –9% for DA1 in the cost of capital in the experimental economic system.

References

Arrow, J. K. (1962). Economic Welfare and the Allocation of Resources for Invention. *In NBER The Rate and Direction of Inventive Activity: Economic and Social Factors. Princeton Univ. Press (for the NBER).*

Coase, R. H. (1946). The marginal cost controversy. *Economica, 13*(51), 169–182.

Kahneman, D., & Tversky, A. (1979). Prospect theory: An analysis of decision under risk. *Econometrica, 47*, 263–292.

Lamoreaux, N. R., & Sokoloff, K. L. (1999). Inventive activity and the market for technology in the United States, 1840–1920. NBER Working Paper.

Lamoreaux, N. R. & Sokoloff, K. L. (2001). Market Trade in Patents and the Rise of a Class of Specialized Inventors in the 19th-Century United States. *The American Economic Review*, 91(2, Papers and Proceedings of the Hundred Thirteenth Annual Meeting of the American Economic Association), pp. 39–44.

Smith, V. L. (2002). Banquet Speech, Nobel Dinner. *Nobel Foundation*.

Ullberg, E. (2009). From personal to impersonal exchange in ideas – Experimental study of trade in organized markets for patents. KTH TRITA-TEC-PHD 09–006, 180.

Ullberg, E. (2010a). Coordination of inventive and innovative activities through markets with prices: An experimental study of patent markets with transparent prices. CESIS Working Paper Series #231.

Ullberg, E. (2010b). Du passage d'un système d'échanges "personnels" à un système d'échanges "impersonnels" des idées: une étude expérimentale (Encadré 3). In D. Guellec, T. Madiès, and J.-C. Prager (Eds.), *Les marchés des brevets dans l'économie de la connaissance, Report for the Economic Counsel of the Prime Minister, Paris*.

Ullberg, E. (2010c). The problem of trading patents in organized markets: A dynamic experimental microeconomic system model and informal price theory. CESIS Working Paper Series #229.

Chapter 4
Coordinating Inventing and Innovating Through Markets in Patents with Prices: Experimental Study of Institutional Behavior and Price Signals

4.1 Introduction[1]

In this chapter, I wish to examine the coordination of inventive, intermediary (trading), and innovative activities through organized markets for contracts on patents with transparent prices. In particular, the willingness to (endogenously) search for new technology within a given search space of (exogenously) induced values is evaluated under different institutions and presumed patent validity in an experimental economic study.

The standard example of such coordination typically does not take place through a negotiation process with transparent prices. This analysis goes beyond the gains from trade due to comparative advantage (Heckscher 1919; Ohlin 1924), static economies of scale (Krugman 1980), and the Schumpeterian world (1942, 1934) in which integrated markets increase the incentives for and the benefits from innovation (a market *size* argument). Such analysis takes a three-phase approach: innovation, rewards, and technology becoming public. The analysis leads to an economy with a permanently higher growth rate (Krugman 1990). The traditional approach follows a "temporary monopoly" discussion on the innovations (products) and invention and innovation are integrated in the same hierarchy.

In this model, invention and innovation are not integrated in the same hierarchy but through specialized agents coordinating their activities through a market with prices. This chapter expands on previous chapters on trade in organized markets for patents with prices and dynamic gains. In Chap. 1, an informal theory of prices for linear contracts on patents was outlined and a dynamic economic system was designed to be useful for experimental investigations (Ullberg 2010c). In Chap. 2,

[1] I am grateful to the Saving's Banks Research Foundation (Sparbankernas Forskningsstiftelse), Dir. Rodrigues, and Royal Institute of Technology, Björn Hårsman for research support which have helped to shape the research agenda. See thesis by Ullberg (2009). In particular, I also would like to thank Vernon Smith, Stephen Rassenti, and David Porter as ICES-GMU for their invaluable input on economic system design and economic experiments.

E. Ullberg, *Trade in Ideas: Performance and Behavioral Properties of Markets in Patents*, Innovation, Technology, and Knowledge Management 13,
DOI 10.1007/978-1-4614-1272-4_4, © Springer Science+Business Media, LLC 2012

prices and gains from trade were investigated and the proposed price theory was tested for three demand-side bidding institutions under two levels of patent validity using a laboratory experiment (Ullberg 2010b). Dynamic gains from trade were also studied as the propensity to "split" contracts on patents in two technical fields of use, pricing, and increasing the use of an *exogenous* technology with uncertain values.

In this chapter, I wish to expand the study by introducing *endogenous* technology search for technology of different values in the dynamic microeconomic system where only the boundaries of search of technology are defined. Such search limits can be said to represent the technological research capabilities of a typical firm. The design uses Smith's (1982) microeconomic static system description[2] with extensions to include a dynamic system with a secondary market and a legal patent environment in a three-period game with multiple system states as in the previous studies.

The key point of interest is the *search patterns* that can be observed using different institutions and patent validity. The attempt is to build a "bridge to reality"[3] when impersonal exchange in patents takes place in organized markets, by staying true to the principles of the patent system of *market exchange and social exchange.*

It is my contention that the price signals and trade in the patent rights based on publicly disclosed technology area information and demand-side bidding market institutions are important aspects inherent to the value of the patent system in the analysis of social exchange that takes place for the development of economically useful technology, specialization, and growth (through a complex coordination of publicly disclosed technology through the patent system and public prices through a market exchange).

I will first give a background in terms of the model used (the trading system in Sect. 4.2) and then a discussion of two search processed studied – one with a simple and one more complex "endogenous" search space (Sect. 4.3). The details of the economic environment, institutions, and software used are then presented (Sect. 4.4) followed by an overview of the experimental design and results (Sect. 4.5). Empirical results and initial hypotheses on coordination and willingness to search (WTS) are tested in a comparative study between the institutions and patent validity in the 3×2 design (Sect. 4.6). Finally, conclusions are briefly discussed and a proposition to address perceived deficiencies of the market economic system of capital investments in new products (innovations) is addressed (Sect. 4.7). Some of the policy implications presented in Sect. 4.7 and in the thesis are elaborated on – in summary form – Ullberg (2010a).

---

[2] The system description $S = (e, I)$ is expanded to specify technology as a knowledge discovery activity (inventing patentable subject matter) part of the economic environment (e) and a legal environment (the patent system) as part of institutions (I). p. 924 "… if learning is to be part of the economic process, then one must specify agent preferences and technology in terms of learning (or sampling or discovery) activities. In this case, the fixed environment would specify the limitations and search opportunities for altering tastes and knowledge in an economy with changeable tastes and resources."

[3] The challenge that remains. Ref to : Krugman (1990).

4.2 The Trading System

Intellectual property has always, it seems, been at the center of economic activity. It has been kept private through many different means, such as trade secrets, but through the patent system technical knowledge has become *impersonally* tradable [see Ullberg (2010c) for a more detailed discussion on the economic history of the patent system]. The technical knowledge remains private through the patent claims, but the information about what is invented becomes public through the patent disclosure. Today, the patent system has become "inseparable" from economic activity and is used in very sophisticated ways to protect market access. The rights allow the holder to exclude and/or transfer or license the rights, creating a *market* in ideas. Traditionally, this market can be characterized as a *personal* exchange market between individuals, firms, and even nations without any organized way of pricing the patents (comparable to that of a stock market for shares). However, during the last two decades, the patent licensing market has dramatically grown in importance with increased specialization between actors inventing and innovating as a result. Also, specialized patent portfolio companies, "Traders," have appeared. This was probably triggered by a "pro-patent" policy shift, particularly in the USA in 1982, when the benefit of the doubt was given to the Inventor, the agent taking the risks. This has had the effect that patent validity – the chance that patents challenged in courts are upheld – has increased (in the USA) from 38% in 1982 to 93% around 1999. The patent has become an asset comparable to a physical asset – presumed highly valid and therefore tradable.[4]

However, this exchange rarely results in prices that are public, strongly "covering" any public *price signals* to Inventors, Traders, and Innovators about the value of certain technology areas. Some specialist firms do collect information on patent licenses and there are general "rules of thumb" for his information, shedding some light on the willingness to pay. However, to date no organized market with prices exists, although many efforts are currently being made in this direction.[5] A reason for this inability to trade impersonally was suggested in the results of the previous experiment (Ullberg 2010b). Both buyers and sellers are uncertain whether public prices would be to their advantage. The experiment shows that impersonal exchange can take place using a linear contract and two-dimensional demand-side bidding. In this study, I use the same design markets for *impersonal* exchange in patents with *public* prices as in the previous experiment, providing the basis for signaling of the value. Inventors are given a "search space" that represents discrete patentable technology areas. Each area has an induced uncertain value for the buyers. The principle implemented here is thus coordination of investments in technology areas through market price signals.

[4] It is illegal to license a patent that one suspects to be invalid, putting an often conciderable cost of "prelicense validation" on the sellers when presumed validity is low.

[5] There are initiatives of auctions such as bancruptcy auctions, government initiatives (Japan), and private organizations (USA), but not yet an organized market with prices.

4.3 Isolation of the Problem

The first objective of these experiments was to determine whether agents would be able to *search* and identify the highest value technology when agents were faced with prices determined in competitive demand-side bidding markets by other agents. The previously cited experiment focused solely on prices of the contract and dynamic trade gains based on multiple uses, given uncertain values for *one* (exogenous) technology. To search among different technology areas for a *better* area, given by higher value for the Users, adds uncertainty to the knowledge development process and demands a certain level of risk-*taking* (entrepreneurial activity) in the breadth of searching.

The search results are not *private* information when using patents, but disclosed in the disclosures section and published by the patent system, thus making the *information* on the appropriated technology *public*. Making your intention (technical area), *public* is the price you pay to the society, i.e., other Inventors, Users, etc., in exchange for a grant of a time-limited *private* right to exclude others from trading that information. The disclosures are in a sense of a "patent product" description. It is this information that is added to the experimental design in this second experiment. The search also demands prices that truly reveal any *differences* in values between technology areas. Consequently, it is an open question to what extent this additional entrepreneurial risk-taking will take place to search the "technology space," a *necessary* condition for coordination through prices.

All technologies have some value; thus, there is the behavioral preference expressed in the prospect theory of "loss aversion," which should limit search for new technology on the margin. There is some "gambling" involved in trying to find a better technology value and an "insurance" in continuing to use the same technology area. The risk-taking behavior may thus be affected by incentives given by the different institutional arrangements to price a new value (convergence rate, ability to differentiate between blocking and investment value, etc.).

The proper integration of information and rules in the institutions tested are therefore critical to integrate privately held information on the value of a given technology. Only through a process giving incentives to broad search can form common expectations that result in efficient outcomes. The level of "willingness to search" (WTS) will thus be a determinant of gains from trade through coordination of public information on the location of the technology area, with public prices and private rights.

If agents are not observed to search exhaustively, this supports risk-averse behavior induced by the impersonal *market* institutions and the *social* exchange of information [everyone is informed of the price (value) of an area whether the "Inventor" benefitted (high value) or not (low value) from the effort]. Also common initial expectations developed during the experiments (trained subjects are used) may be a limitation to search (the likelihood of finding a "higher" value than the highest found can be estimated). It is thus the principle of disclosure that is implemented in the dynamic model to make price signaling of fruitful technology areas possible.

The second objective was to investigate any differences in *gains* from trade through price coordination between the design market institutions and patent validity used. The incentives given by the rules and validity of patents influence prices and therefore may influence coordination efforts and convergence. Entrepreneurial activities carried out in the same hierarchy tend to create new knowledge on the margin, to reduce risk-taking.[6] A comparison between the institution similar to today's *personal* exchange and the others with richer language for demand-side bidding would reveal differences in gains from trade coming from an *impersonal* exchange. This question thus addresses whether there is a change in systemic risk-taking between the different means of economically organizing inventive activities to a hierarchy or specialized agents who coordinate through a market with prices. Will subjects interacting through a market with prices socially behave with more risk-taking, i.e., will the risks be shared in a more effective way? This is a systemic change that contains the *social exchange* aspect of the patent system.[7]

To create an environment with uncertain values that could be explored during a single session, initially nine technology areas were offered but only three had any profitable values. This turned out to be too a simple search problem, and nine random values gradually spanning from preference for investing to preference in blocking were chosen. The second "search space" turned out to be much more challenging, revealing preferences for investing and blocking for different institutions as well as multiple convergence end-states.

4.4 The Economic System: Details of the Institutional, Economic, and Legal Environment

The trading procedure employed in this study is a modified version of a specially built computerized trading system[8] for linear contracts. The software incorporates *endogenous* choice of technology area, the three institutional mechanisms tested (the primary markets), individual screens for the Inventor (Role 1), the Trader (Role 2), and the two types of Innovator roles (Role 3A and Role 3B), as well as a fixed price double-auction mechanism (the secondary market) used to re-trade the already negotiated contracts in period 1, in periods 2 and 3. There are three periods in each round. The trading mechanisms and patent validities are identical to those of the experiment for prices and gains from trade with *exogenous* technology values. The participant's screen displays were different for the different roles. An instruction set

[6] This has also been the development of the patent system, from rather high demands to get a patent to gradually lower demands for patentability, making basic work efforts patentable.

[7] This question has a parallel in key management questions (hierarchy) where the information-sharing inside the organization typically results in a better outcome for the group.

[8] The software is a modified version of the software used for Experiment 1 on prices and dynamic gains from trade through multiple use of patents.

used during the experiment allowed participants to learn the interface more quickly by providing each participant with a detailed explanation of the different areas, boxes, and information on their screen and what each role could do.

4.4.1 The Flow of the Experiment

Figure 4.1 gives an overview of the state diagram of the endogenous experimental flow in the experiment.

The experimental flow has three general phases which are repeated in each round: Inventing (1), Trading (2) (3), and Using (6) (4) (5). Inventing, primary market Trading, and Using take place in period 1, and secondary market Trading (5) and Using take place in periods 2 and 3. The actual flow through the states is executed by the decisions of the participants. The Inventor is first asked to select a "technology focus" which is represented by nine "radio buttons" on a "technology map," representing the technical field of the Inventor (1). Each selection has a *different*, initially unknown, set of induced (private) values. The Inventor is then asked to create a linear contract of "standard" or "quality" (1). The *private* values in a round are given for the three periods when the contract has been issued by the Inventor. The *private* values for the Trader and Users, and the *public* information on the technology focus and contract quality are given when the Inventor starts the auction by "listing" the contract with a first ask.

The contract is predefined as a license on an invention with "technology focus" "AB," useful for producing products of type "A" and "B." A quality contract can be split into one contract with focus "A" only and another contract with focus "B" only. The Trader is the agent who can split the AB contract into the A and B contracts; a quality contract thus allows the Trader to participate in the bidding process (2). If the Inventor decides to Invest in a standard contract, which cannot be split, the Trader is left out of the bidding (2) for that round. If the Trader wins the AB contract, and then subsequently splits it, the A and B contracts are sold in sequence to the Users of Role 3A and Role 3B, respectively, with the Trader now being the seller and the Users the buyers (3). A User of type A can only produce products of type A (their "product mix"), and vice versa for type B Users. If a standard contract is sold, all Users can participate, but there will be only one winner and one participant (A *or* B type) who can use the contract. If the contract is split, there are two participants (A *and* B type Users) who can use the two contracts and the Trader who used the AB type to split. As each bid–ask is entered, that information is *publicly* provided to all agents. The contract price (fixed plus royalty) is then a *price signal* for the value of the technology area and the contract quality to the Trader and Users (ultimately the consumers of the products A/B).

The (binding) contract(s) sold thus eventually end(s) up in the "portfolio" of one User (with product mix A or B) or two Users (one with product mix A and one with product mix B) and a Trader (who holds the AB contract issued by the Inventor).

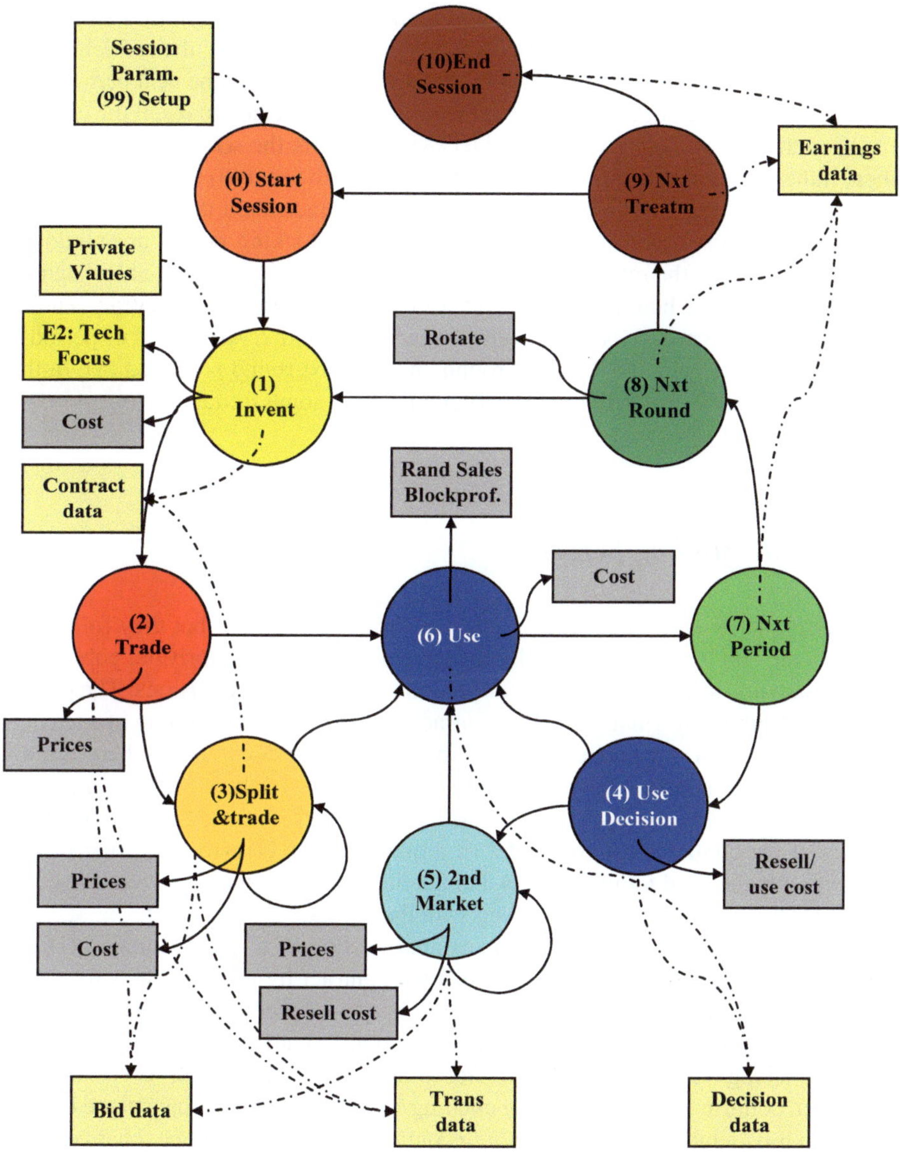

Fig. 4.1 State diagram of the endogenous experimental flow in the experiment

The User(s) holding the contract(s) are now asked (6) to either "Invest," which produces a profit based on increase in sales of the new more competitive product(s) A and/ or B invested in, or to "Block" which produces a profit based on the existing sales of the product(s) based on existing technology the firm has been endowed with, "insuring" the firm's market access against competition based on technology.

The sales and blocking profits are randomly drawn from a uniform distribution[9] in their respective induced value range displayed to the participant, after which period earnings are calculated for period 1 and the experiment moves to the next period (7).

At the beginning of periods 2 and 3, the possibility to decide to resell the contract (or keep the contract) is given the holder (4) who starts the secondary market (5); then, if the contract is resold, the new holder is asked to use it (6) (cannot resell until next period). If the holder decides to keep the contract, then the holder is asked how to use it this period (6), after which the sales/blocking profits are randomly drawn anew within the respective value range and period earnings are calculated for all participants who held (bought/resold) or issued (in period 1) a contract, ending period 2. Period 3 proceeds in the same manner as period 2. At the end of period 3, total earnings are calculated for the round and the next round is started (8). In this round, roles may be rotated among participants with some frequency (e.g., every 2 or 3 rounds).

4.4.2 The Trading Screens

The participant's screens are rather complex "trading screens," but they follow the same theme and logic as described. The screens sections cover private, public, and earnings/status information, useful for participants in making decisions. The private values and public information on the contract are given at the beginning of each new round. The values are given in ranges for the three periods. Thus, there are six ranges (three for sales increase if contract is invested in and three for blocking profit if contract is used to Block). Once Role 1 has decided the technology focus and quality of the contract, Roles 3 and 2 (if a quality contract) will have their private values displayed. The *public* validity and quality, as well as the the *private* discount rate for the contract, are also displayed with the values.

In the first mechanism, a two-dimensional version of a double-auction (DA1), participants enters a price to buy (or sell) the contract by entering their fixed *and* royalty bid (offer) and then clicking on the button labeled "Send Bid" (or "Send Offer"). Participants can also accept any other participant's bid to buy (or offer to sell) by clicking on the button labeled "Accept Bid" (or "Accept Offer"). The acceptor must then confirm the acceptance by clicking "YES" (or "NO" to cancel the bid) on a pop-up box, which results in a *binding* contract to be formed; the exchange information is recorded in the public market information section on the interface. The auction is started by the seller (Inventor/Trader) with the first ask.[10] All past transactions are also listed with the last transaction first as a memory for the participants regarding negotiated prices for different markets (primary market AB, A, B, technology focus,

[9] Other distributions were discussed in Chapter 1 (a Poisson distribution).

[10] This is different from a typical one-dimensional double-auction where the auction is started when the contract is listed. The buyer or seller can send in the first bid–ask.

resell price, validity, and quality). The participants are given calculation tools to automatically calculate expected profits for rational expectations (mid-point value in range) and whether investing or blocking is decided.

Price quotes must reduce the bid–ask spread in one or both dimensions to be accepted. A buyer can ameliorate a bid, for example, by increasing the fixed bid without increasing the royalty bid. The five highest bids and lowest asks are displayed in a ranked order visible for all participants. Only the highest bid and the lowest ask are open for acceptance. Price quotes that violate this rule are rejected. Identical bids are thus rejected. The auction is started by the seller (a Role 1 or Role 2) and ends when the bid and ask meet in both dimensions, when an "accept" is made, or when the auction times out, resulting in a "no trade" for that contract. In the case of "no trade," the seller must bear any costs associated with the creation of the contract (Role 1) or loss of fixed and royalty fees to cover cost against obligations from contract bought (Role 2). Buyers are not affected by "no sale." The timeout function has two timers to speed up the bidding process: one for the maximum total time for the auction (usually 120 s) and one for the maximum time between bids (usually 20 s).

The second mechanism, a double-auction with reservation value on the fixed fee (DA2), is identical to DA1, except when it comes to the amelioration rules. The seller can only *increase* the fixed component, i.e., the initial fixed quote is a *minimum*, not a maximum. The buyer can only *decrease* the fixed bid, i.e., their initial fixed quote is a *maximum*, not a minimum. This institution thus gives the seller the privilege of setting a binding minimum price on the fixed fee, which is a de facto reservation value. The royalty bid works the same way as in DA1.

In the third mechanism, repeated posted-offer, (PO), or a "manual" Dutch Clock auction, the seller enters a price quote in exactly the same way as in DA1. However, the buyers are limited in bidding space to a simple accept or reject of this offer by clicking on the button labeled "Accept Bid" or "Reject Bid." The information on who or how many have accepted or rejected the offer remains private. If all buyers reject the offer during a bidding round, then the seller can ameliorate the offer by reducing the quote in one or both bidding dimensions, just like in DA1, but here at a cost for each ameliorated offer. The bid–ask gap is unknown to both buyers and sellers until the quote is accepted by at least one buyer, at which time the binding contract is formed. The auction ends when the *first* buyer accepts the quote in the bidding round or there is a timeout. This mechanism could perhaps be called "manual Dutch Clock in two dimensions," where the "clock ticks" are provided by the seller's reduction in quote in every bidding round. Only the past offers are listed in a ranked order in the public market information.

The secondary market (5) is a standard fixed price double-auction (FP-DA) common in asset experiments where the seller enters a fixed price quote for the contract in its portfolio, bought in a previous period. The contract terms (fixed, royalty) agreed upon in the primary market (2) or (3) are not re-negotiated and the contract is transferred "as is." Price quotes can be positive (seller gets money from the buyer) as well as negative (seller pays money to the buyer). This mechanism allows for already negotiated terms to be compensated for to a level acceptable to a buyer. The buyer (sellers) can also accept the highest standing ask (bid), as in DA-1.

Contracts can be resold in periods 2 and 3. The auction is started as the seller decides to resell (4). The first quote can thus come from either the seller or a buyer. The quotes have to reduce the bid–ask spread. The auction ends when the quotes meet, an accept is made, or the auctioned is timed out. There is a timeout for the total auction time (90 s) and for the maximum time between quotes (10 s).

Trading occurs over a maximum of 30 rounds, each having three periods, lasting approximately 1–5 min each per round.

4.4.3 Subject Payments, Endowments, and Special "Rules of the Game"

At the beginning of the experiment, initial roles are assigned to the participants. There is one Inventor (Role 1), two Traders (Role 2), and 6–8 Users (Role 3), with half in "industry A" capable of producing product A, and half in "industry B" capable of producing product B (if an odd number of users is involved, there is one less in industry B). Each participant is endowed with a capital of experimental money at the beginning of a treatment. The purpose of the capital is to introduce "bankruptcy laws" and "loss aversion." The participants' earnings are decided by accumulated gains (losses) via contract issuing, splitting, and using contracts held in portfolio during each period. At the end of the experiment, participants are paid a weighted sum of accumulated earnings in each role plus an hourly fixed fee (for keeping the capital positive) and a fixed show-up fee (for showing up on time), not counting the experimental money capital endowment. The exchange rate is decided at the end of the experiment (partly due to time constraints on the number of rounds in each session), and "converged" to 0.1 for Role 1, 1 for Role 2, and 0.5 for Role 3, in order to make all roles approximately equally profitable (which turned out to be a difficult task for most agents).

If the accumulated losses deplete the capital, a one round grace period is given to the participant ("Chap. 11"); if the capital is still negative at the end of the next period, the participant is declared "bankrupt" and cannot participate further in that session. If this occurs, both the hourly fee and the show-up fees are lost (there is no payment for the session). This seemingly harsh measure was introduced to stop overbidding by many participants in the initial sessions. Participants are informed in the instructions about the linear nature of the contract, the decisions and activities each role can undertake, the uncertain nature of the sales and blocking profits generated by using the contracts, and the "bankruptcy rules." They can keep the instructions with explanations of the trading screens and experimental flow during the whole session. They are not informed about the distribution of values among participants (which is a linearly increasing function). At each session, the technology map is randomly rearranged so that there is no prior expectation to the value locations. However, common expectations develop regarding "possible" values for the technology.

Invalidated contracts annul payment obligations (typical practice in real world). If a contract is invalidated, this information is immediately displayed in the market info box.

The number of participants varies between 9 and 11 in all sessions, changing the competitiveness of the demand-side between sessions. Each session is reported separately and all data can be related to an individual.

4.5 Experimental Design, Design Parameters, Performance, and the Sequence of the Experiment

I report findings from 14 experimental sessions using the design parameters listed in Table 4.1. A 3×2 design was used for the study of coordination and dynamic gains from trade under three primary market institutions and two levels of presumed patent validity. Six designs were run, exploring the coordination efforts in each combination of institutional and legal environment.

The initial sessions (1 to about 3.2) had simpler value sets and served as pilot/ data sessions in their own rights. From session 4.1, more complex data sets with more relevant search space were used to better test the robustness of and differences between the institutions. The last session's values spanned from preference to Block to preference to Invest, adding this aspect of the coordination problem to the test. The induced investing and blocking value ranges were positive, exploring the linear contract for positive prices as in the previous experiment. The values for the investment and blocking values were given in overlapping ranges, linearly increasing for the buyers to create a competitive bidding environment with only one "high-value" bidder. The Inventor was given a broad range spanning over the ranges of all the buyers, a typical condition for Inventors. The Traders were also given the full range, but spanning over each "industry's" ranges. This was based on the presumption that Traders typically have better information than Inventors, being closer to the product market and having an "industry" view. The Users had the most narrow (precise) ranges of values being in the market using the technology. A fixed cost of capital, specific to each role type, was used to discount the values over the three periods in each round. See Fig. 4.2 for an outline of the values used. The values show the midpoints of the range given to the highest value User.

The first session was run both as a pilot, perfecting the design, and a data session. Subjects were recruited from the same pool that had experience from the previous experiment with prices and exogenous technology. This made training time minimal and the markets worked immediately in the new environment with the additional endogenous choice for technology. Data for technology focus were recorded in addition to the previously already-recorded quality decisions, bids, time series, prices, gains, and other dynamic system parameters.

Experiments with high validity are reported first (design 1–3), by institution and market, followed by the experiments with low validity (design 4–6).

Table 4.1 Experiments by institution and legal environment

Design	Institutional and legal environment		Competition type	Contract values			
	Institution	Validity (%)	A and B Users	(contracts in session)[a]	Values.Set	Experiments	Data (rounds)
1	DA-1	93	(4+3)	3 AB, 3 A, 3 B	Values9.1	12.1	41
		93	(4+4)	3 AB, 3 A, 3 B	Values11.2	3.1	11
		93	(4+4)	9 AB, #5 low	Values13.1	3.2	20
		93	(4+3)	9 AB	Values14.2	4.1	30
		93	(3+3)	9 AB	Values15.3	7.1	30
2	DA-2	93	(4+4)	3 AB, 3 A, 3 B	Values10.2	2.2	22
		93	(4+4)	9 AB	Values14.3	5.1	18
		93	(4+4)	9 AB	Values15.2	8.1	31
3	PO	93	(3+3)	9 AB	Values14.4	6.1	22
		93	(3+3)	9 AB	Values15.4	8.2	30
4	DA-1	38	(4+3)	9 AB	Values16.1	9.1	26
5	DA-2	38	(4+3)	9 AB	Values16.2	9.2	30
6	PO	38	(4+3)	9 AB	Values16.3	10.2	37
							348

Cost structure:
Issue cost for standard control 1 and quality 2; transaction cost=1, COGS=65%, investment=1, patent renewal=0
Cost of capital: 30% for Inventors, 5% for Traders, and 10% for Users
[a]The values are given in a range (a,b) to the subjects and a random value is drawn for the realization of the value using a uniform distribution. Different value sets were used to randomly change the value for the technology areas between sessions. From session 3.2, a low value was always given for #5 to avoid constant use of "center" position

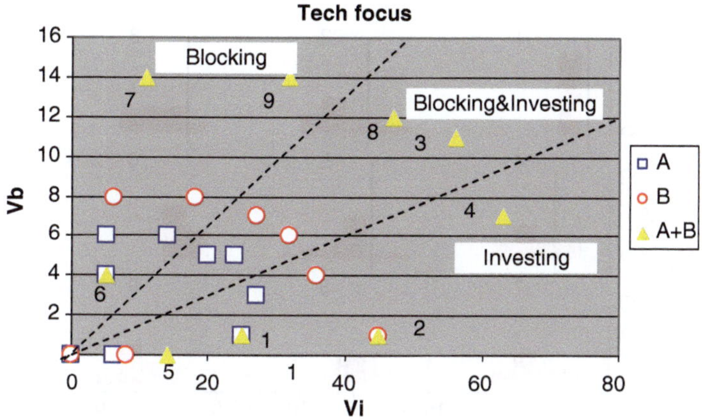

Fig. 4.2 Induced values for sessions 4.1–10.2 for technology areas 1–9. The technology values vary in preference to investing or blocking and NPV, with 4 the highest. For sessions 1.1–3.2, a simpler value set was used

4.5.1 Coordination and Gains from Trade Under High Validity

4.5.1.1 Some Initial Observed Coordination Dynamics in the Primary Markets

The distribution of prices for the different technology areas 1–9 are visualized in Figs. 4.3 and 4.4 for the market mechanism DA1. In this "9 commodity market" in two dimensions with uncertain values, pricing patterns clearly suggests that the institution appears to generate distinguishable prices compared to predicted prices, although with volatility, especially in royalty.

4.5.1.2 Coordination by Institution

The first experiments used a value set consisting of 3 AB, 3 A, and 3 B contracts, where the AB contracts could be split into one A and one B contract. The initial concern was that the price signals in this already complex "9 commodity" experimental market with linear contracts would not be clear enough for coordination decisions.[11] The values are called 1–9 in the figures and were assigned to the radio buttons with the AB contract on the diagonal, the A contract above and the B contracts below the diagonal. The values were randomly assigned in each category. The search pattern and convergence over time for Experiment 1.1 is shown in Fig. 4.5.

[11] In the experimental auction literature, often 3–4 rounds are needed to create common expectations and reach a theoretical price or stable trading pattern in a *one*-commodity market.

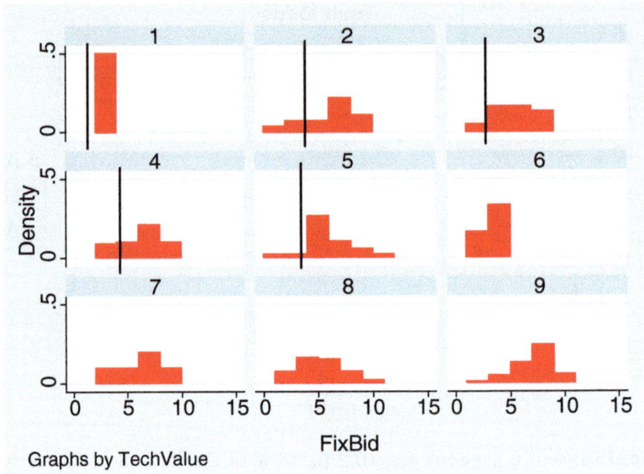

Fig. 4.3 Fixed fee prices for technology areas under DA1. Average prices are indicated as *solid lines* and predictions as *dashed lines*

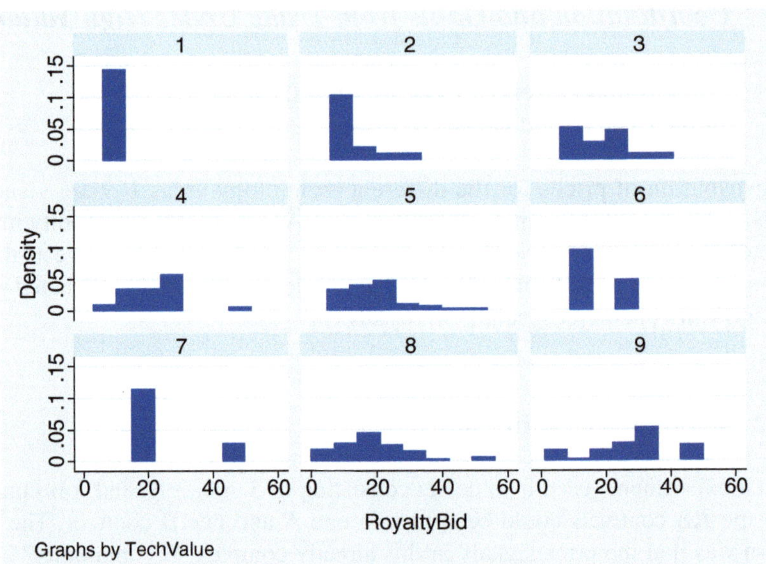

Fig. 4.4 Royalty (%) prices for technology areas under DA1. Average prices are indicated as *solid lines* and predictions as *dashed lines*

Subjects tried 8 of 9 technology areas, early discarding the lower valued A (#1,6,7) and B (#2,8,9) contracts, then tested out the three AB contracts (#3,4,5) and appeared to settle for area #4, which is the area with the highest NPV. In the process, 4–6 contracts were priced for *each* of the areas 3–5, giving *prices* a chance to converge (subjects had the transaction history on their screens). Average prices (fix, royalty) for #3 were (6, 22), for #4 (7.5, 25), and for #5 (6.6, 18.4); thus the average price

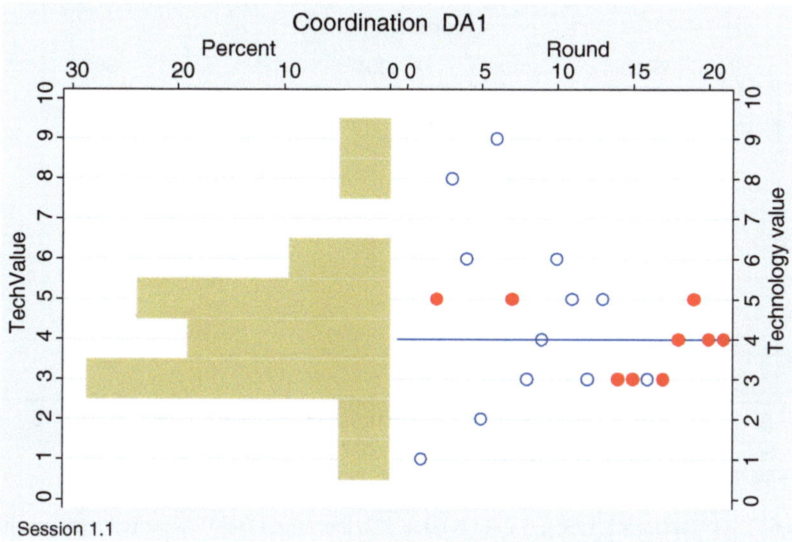

Fig. 4.5 Convergence in session 1 for DA1. The technology areas are shown on the *y*-axis with #4 having the highest net present value. *Filled dots* indicate quality and *hollow* standard contracts

signal indicated that #4 is the most valuable for the Users, irrespective of its use (Invest or Block). Another way to look at convergence is by frequency. A high frequency would indicate the perceived usefulness for that tech area under rational expectations, a measure used in all of the experiments hereafter. The contract type issued was initially the cheaper standard contract and later a splitable quality contract, indicating that a narrowing search was first done at a lower cost before more valuable and costly contracts were invested in. In the following sessions, 2.1, 3.1, and 3.2, limited search was observed, suggesting common expectations had formed in the first session about the AB contracts being more profitable than the A and B contracts. Figure 4.6 charts these sessions.

What appeared to be a tendency to select the geometric center of the technology map (#5) was observed. To avoid what appeared to be a selection bias for the center of the map, or possibly too little variation in the AB contract values for price differentiation, a low value was given for area #5 in session 3.2 to increase the incentives for search. This resulted in a renewed search, still among the AB contracts, and #4 was discovered and received the most "hits." However, still #5 received many "hits."

These initial and convincing results of coordination through linear prices motivated a change in the value set to broaden the search space from 3 to 9 AB contracts, not using the A- and B-only contracts. The idea was to use a range of values spanning from predominantly investment values to "dual" Invest/Block values to predominantly blocking value. *Competing* price convergences would be possible (given individual tastes for risk-taking in the search). These value sets would then provide a difficult "robustness test" of the coordination for different institutions, where preferences would interact with willingness to take risks (gamble) in the search.

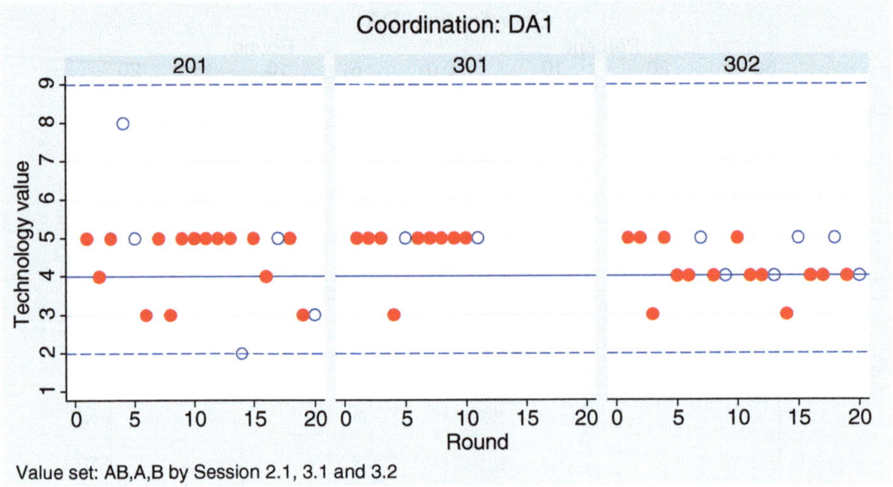

Fig. 4.6 DA1: sessions 2.1, 3.1, 3.2. In session 3.2, the values for #5 were reduced, resulting in increased search among the three AB contracts (#3, #4, and #5)

The following sessions were thus considerably more complex in nature, with only AB contracts resulting in more complex search and convergence patterns.

Comments from subjects suggested that once "OK" contract values had been found, they tended to stay with those, since "I might end up with a low-value contract when it is my turn in Role 1 (Inventor)." To remedy this "high-risk low-return" problem, the roles were rotated first every three rounds, then every two rounds (to give "two chances"). When rotated every round, the risk-averse behavior was clear. Risk-averse behavior appeared to be a common behavior. Only a few people responded differently by searching for a new area when others did not. When asked after the session why they searched so extensively, one person said "I always want to try some new thing. Perhaps there is a good value there." Another said "If I try something new, then the others will find out whether it is good or not, and then when it is my turn again, I will benefit also." These comments are consistent with risk-taking and *social* exchange through *impersonal* markets. In the patent system, a motivational *principle* is disclosure of what is invented so that others may benefit, not only after the patented technology has expired, but also as an input for further research in a particular area. This principle appears to be at work in a social context with some of these subjects. *Searching* may thus be a social activity, not just an individual activity, given a patent system.[12] However, entrepreneurial people were "a rare flower."[13]

[12] Compare Arrow's definition of rationality: "Rationality and knowledge of rationality is a social and not only an individual phenomonon" Arrow (1986).

[13] Discussing the problem of limited willingness to search with Prof. Vernon Smith, he thought that these particular subjects just might not be willing to jump in there, adding that "entrepreneurial activity is a rare flower."

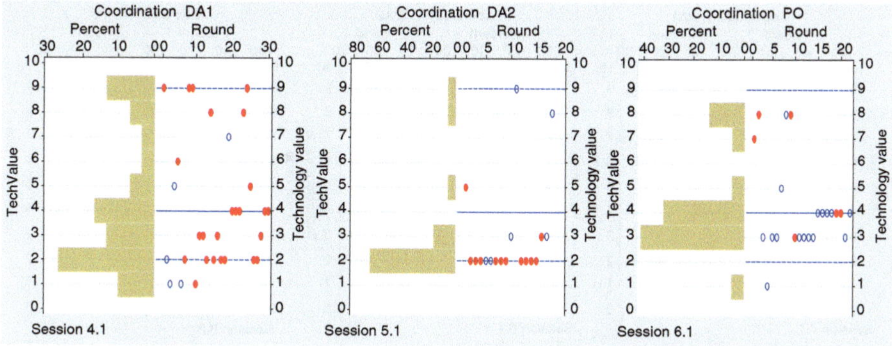

Fig. 4.7 Session 4.1 (DA1), 5.1 (DA2), and 6.1 (PO). Dashed lines indicate highest investment (#2) and blocking (#9) value. Highest NPV is at #4. Circles are standard and filled circles quality

In Experiments 4.1, 5.1, and 6.1, charted in Fig. 4.7, the more complex value set was used across the three institutions. For tech areas #1–4, the investment values dominated and for #6–9 the blocking value dominated, with #5 having low values for both. A different search pattern appeared for the mechanisms, clearly distinguishing the institutions in propensity with respect to" (1) WTS, (2) convergence to any given technology area, and (3) dynamic gains from trade (measured as propensity to split). For the DA1, the whole tech area was searched at least once, low-value contracts were quickly rejected, and high-value contracts were revisited. These high-value contracts were characterized by: focus on investing (#2), focus on blocking (#9); and dual focus #4. The most used contract, including the dynamic gains from split was #4 (highest NPV). The last two contracts were also #4. This indicates that DA1 appeared, in this session, to discover the three high-value contracts as well as to "converge" to some extent on the highest NPV. We also see that contracts of some value are revisited 2–3 times before abandoned, presumably in order to get a clearer price signal (enough information to form common probability expectations). For DA2 and PO, there were less search and price discovery, even for valuable contracts. In the case of DA2, where a reservation value is set on the fixed by the seller, the contracts with high investment value appear to be most attractive (#2). Such contracts allow for trading off some royalty for a fixed. In the case of PO, a more clear focus appears. Coordination converges for the "dual-value" contracts (#3,4), first for #3, and then to #4 (the highest value contract), but with *no* trade gains from split contracts. It is clear that the coordination even in this complex environment appears to converge to the higher values. What appears to differ between the institutions is the *type* of contract values (Invest, Block of both) the search processes converge to. Although this is "noisy," the WTS data appears to be related to the number of demand-side bidding dimensions.

The next set of Experiments 7.1, 8.1, and 8.2 (see Fig. 4.8) had the value set shown in Fig. 4.2. The values systematically covered the value space between investing and blocking with more pronounced blocking values than the previous experiments.

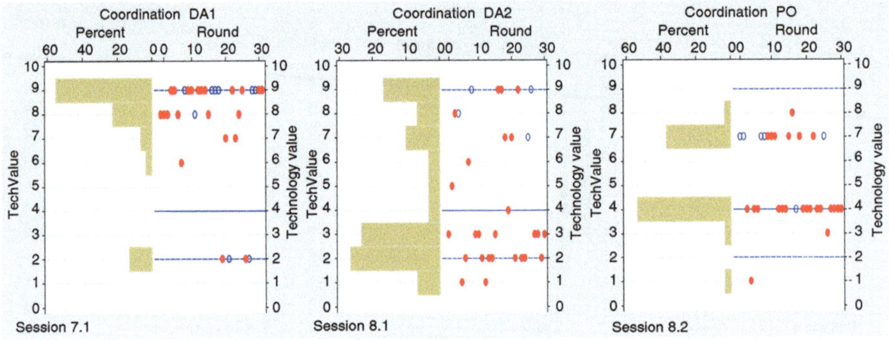

Fig. 4.8 Session 7.1 (DA1), 8.1 (DA2) and 8.2 (PO). Dashed lines indicate highest investment (#2) and blocking (#9) value. Highest NPV is at #4. Circles are standard and filled circles quality

A different search pattern appeared. DA1 converged to #9 and basically stayed put at the high blocking high investment value contracts. Interestingly, area #2 had about the same NPV for investing but almost no blocking value. The contracts were also used mostly to Block. This may suggest a preference for blocking.[14] Again it is one of the high-value contracts where the convergence is strong. In DA2, every technology area was searched with highest frequencies for high investment (#2), blocking (#9), and dual contract (#3). The highest value of #4 was tested but not discovered. The convergence appears to be on #3, which has the highest NPV of the three. For PO, the search pattern is similar to the previous ones, in that it clearly appears to be the most limited. The convergence appears to have two "equilibriums" with one at #7 (which has the highest blocking value, together with #9) and #4, the dual-value contract with highest NPV.

4.5.2 *Coordination and Gains from Trade Under Low Validity*

A last set of experiments covering low validity are chartered in Fig. 4.9. These experiments were run as the last sessions and thus with the most experienced subjects. Clearly, a broader search is taking place. Also, there are few gains from trade coming from split contracts. The similarities between the institutions are also bigger. These two results were also observed in the previous experiments. All end up in coordinating at #2. DA1 basically keeps the focus on #2, whereas DA2 moves from #7 to #2 and PO has the "characteristic" of dual convergence with #2 and #9, as for high validity. The speed of convergence is clearly DA1>DA2>>PO (with possibly two Nash equilibriums). The speed of convergence is of course critical to the efficiency in a technology search process. The general tendency with DA1~DA2>PO appears to remain.

A summary of all six treatments is shown in Fig. 4.10.

---

[14] This is argued in the previous article to be consistent with prospect theory and "loss aversion."

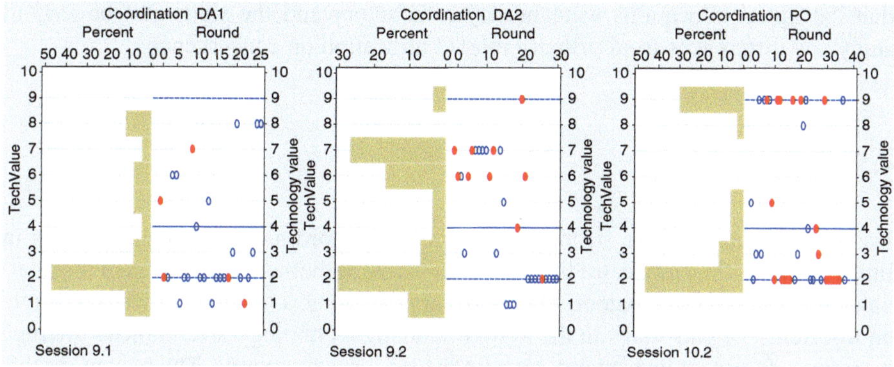

Fig. 4.9 Low validity, sessions 9.1(DA1), 9.2 (DA2), and 10.2 (PO)

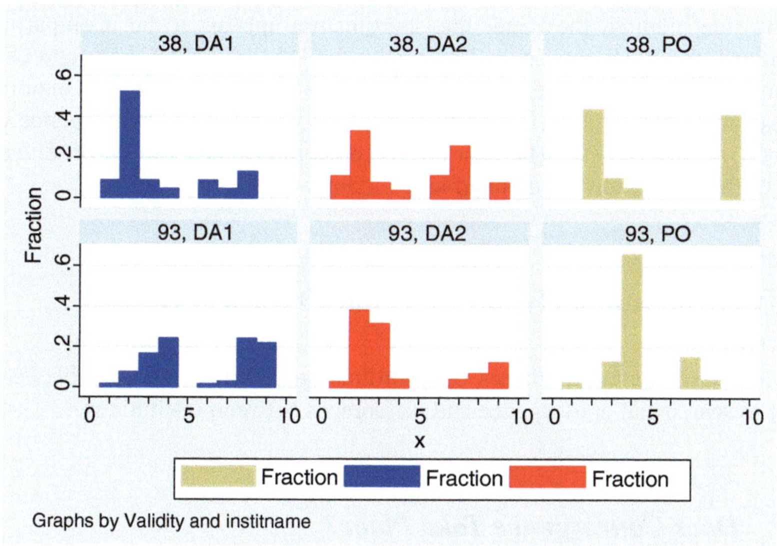

Fig. 4.10 Summary of search patterns for different institutions and validity. Data from #5 are not shown

High validity leads to more searches and an indication of WTS appears to be: DA1 > DA2 > PO.

4.5.3 Secondary Market Usage

The secondary market, used to reallocate contracts in periods 2 and 3, was used only three times for DA1 and one time under a DA2 treatment. The conclusion here is that the initial allocations were mostly satisfactory and the study can be said to study any differences from primary market allocation on convergence.

4.5.4 Summary

I conclude that there are strong indications that coordination through markets in linear pricing of inventive, trading, and innovative activities is supported in the dual-value environment of a patent. There also appears to be a tendency toward multiple equilibriums, or end-states in the system, making technology area choices "swing" between a dominant investment, blocking value, or a dual-value. The reason for this appears to be the investment and blocking values of the contract on patents traded, and the strength of the price signal in the two dimensions, i.e., the institutional characteristics. 1–3 "trials" were made to accept/reject a technology area. The richer the demand-side bidding, the better the coordination appears to be. Comparing the institutions, the WTS appears to be broader for DA1 and DA2 than for PO. DA1 appears to better identify the values (e.g., in 7.1 DA1 discovered #4 as a major value whereas in 8.1 DA2 did not). A tentative ranking in search and convergence would then be DA1 ≥ DA2 > PO. The best convergence thus appears to be observed as a result of the richness in the demand-side bidding.

4.6 Coordination, Hypotheses, and Empirical Results

In formulating some hypotheses on coordination, I will distinguish between the ability to search and convergence and differences between institutions.

4.6.1 Does Convergence Take Place?

Convergence is defined as a repeated selection of a technology area with a frequency greater than 4. With this criteria, there is clearly a convergence in every experiment for every institutional and validity treatment. However, in some cases, in particularly in PO, there are often two tech areas with convergence, one with a high investment value and the other with a high blocking value.

4.6.2 Differences in WTS Between Institutions and Validity

To compare search patterns between institutions, a χ^2 test is used to compare the frequency distributions of technology value searches. The hypothesis is that the institutional rules result in different incentives and different search patterns. See Table 4.2.

Table 4.2 Comparisons between institutions

	Instit			
TechValue	PO	DA1	DA2	Total
93% Validity: sessions 4.1, 5.1, 6.1, 7.1, 8.1, 8.2				
1	2	3	2	7
2	0	13	28	41
3	10	8	23	41
4	52	11	1	64
5	0	1	1	2
6	0	2	0	2
7	12	5	3	20
8	3	19	5	27
9	0	41	9	50
Total	79	103	72	254
Pearson χ^2	0.000***	0.000***		
(Pair wise comparison)		0.000***		
38% Validity: sessions 9.1, 9.2, 10.2				
1	0	2	3	5
2	18	12	9	39
3	4	2	2	8
4	2	1	1	4
5	0	3	0	3
6	0	2	3	5
7	0	1	7	8
8	0	3	0	3
9	17	0	2	19
Total	41	26	27	94
Pearson χ^2	0.000***	0.101		
(Pair wise comparison)		0.000***		

Significance: ***= 99%, **= 97.5% and *= 95%

The conclusion to this is that the search patterns are significantly different for high validity and for low validity, except between DA1 and DA2 for low validity. The differences appear to be that DA1 discovers more precisely the highest valued technology area (4) and highest investment-only (2) and highest blocking-only areas (9). There are 1–3 end-states observed. The frequency "distribution" actually matches fairly well the value distribution of the technology areas. A possible explanation for this is that DA1 allows buyers to bid independently in fixed and royalty; thus, buyers and sellers are able to reach an agreement if one or both values are high (supports the informal price theory). DA2 appears to converge on highest investment-only area (2) and high dual values (3), but with less emphasis on blocking-only areas. A possible explanation here is that since sellers set a reservation value on fixed, a high investment value allows the sellers and buyers to trade off some over-pricing with respect to the blocking value by reducing the royalty. Such behavior would favor contracts with high investment values. This is not possible with blocking-only contracts. A reason why the highest value (4) was not discovered in the session may be that it was only tried one to two times and it appears that three times were needed to accept/reject a technology area (*price* conversion). PO has considerably less search span and appears to stay on initially found high values, often dual values (4, 7). A possible explanation for this is that since the sellers set the offer in two dimensions, buyers are more willing to accept when the values, which are unknown to the sellers, are also matching, i.e., also in two dimensions.

4.6.3 Coordination

Coordination appears to work as technology focus converges on contracts with among the highest blocking and investment values. The different search patterns observed for the institutions suggest, together with the convergence, that there may be multiple end-states of convergence. These are typically on a high blocking value, high investment value, or high blocking and investment value contracts. There is an indication that the coordination is more stable (a single sustained Nash equilibrium) with increased demand-side bidding (DA1 > DA2 > PO).

4.7 Conclusions

Coordination is sustained between invention, trade (intermediation), and innovation through a market with linear prices. In the technology space searched, all institutions converge on the one or two highest values, often resulting in multiple end-states (Nash equilibriums). In terms of WTS broadly and coordinate to the highest technology value area, a tentative ranking can be done between the institutions where DA1 ≥ DA2 > PO. This suggests that *demand-side bidding* in one (DA2) or both dimensions (DA1) are important for gains to trade from coordination through market prices. Since PO is the institution most similar to today's personal exchange, the results indicate that gains from trade in organized markets with prices may come from better allocation of resources for *invention* in a market with prices using demand-side bidding. Search also appears to be motivated by personal and systemic exchange (social exchange motivated by the patent system disclosure principle) as well as impersonal exchange and profits (market institution) as subjects try new areas not only as a one-shot "gamble" but with learning effect for others that may benefit them later ("sowing" in one role and "reaping" in another role or in the same role), whether this behavior is reciprocated or not.

There is a clear unwillingness to continue to search once a "profitable enough" value has been found. This results appears to be consistent with prospect theory (Kahneman and Tversky 1979). The differences between the institutions can then be interpreted as a risk premium with respect to search cost (Diamond 1971) This risk premium is on the price risk, since the difference between the institutions is entirely in the language of demand-side bidding (same contract and same two-dimensional ask from the seller).

Another aspect, possibly worth to be explored further, is the WTS in relation to the endowed (available) capital of the Inventor. This appears to have bearing on Bernoulli's work on a new theory of measurements for risk (1738 (1954)). Here the amount of capital invested is related to the margin of the total capital, making "wealthy" investors act more rationally and "poor" investors willing to accept a lower price since they have more to lose measured as part of their total wealth. This idea is further elaborated on in the conclusions and policy discussion in Chaps. 5 and 6.

I conclude that demand-side bidding in two dimensions reduces the price risk of the institution and is positively correlated to the WTS and thereby correlated to gains from trade with respect to allocation of resources for *invention.*

The model can therefore be seen as an extension of allocation of resources in a hierarchy (Arrow 1962) to such allocation through a market. The model is also an example and extension of the provisions by Smith on his model of microeconomic system where changes in learning can be characterized as discovery activities with some search limits altering knowledge in an economy with changeable resources (Smith 1982, p. 924).

These result rejects, at a level of principle, Robinson's contention (1977, p. 1337), that the problem in the capitalistic system for allocation of resources for innovation is the management of funds and new ideas in the *same* hierarchy, leading to underdevelopment of new ideas. The exchange of the ideas themselves in competitive markets with prices may provide a solution to this problem, as reported in these experimental gains from trade.

References

Arrow, J. K. (1962). Economic Welfare and the Allocation of Resources for Invention. *In NBER The Rate and Direction of Inventive Activity: Economic and Social Factors. Princeton Univ. Press (for the NBER).*

Arrow, K. J. (1986). Rationality of self and others in an economic system. *The Journal of Business, 59*, S385–S399.

Bernoulli, D. (1954). Exposition of a new theory on the measurement of risk. *Econometrica, 22*, 23–36.

Diamond, P. (1971). A model of price adjustment. *Journal of Economic Theory, 3*, 156–168.

Heckscher, E. (1919). The effect of foreign trade on the distribution of income. *Ekonomisk Tidskrift, 21*, 1–32.

Kahneman, D., & Tversky, A. (1979). Prospect theory: An analysis of decision under risk. *Econometrica, 47*, 263–292.

Krugman, P. (1980). Scale economies, product differentiation, and the pattern of trade. In pp. 950–959.

Krugman, P. (1990). Endogenous innovation, international trade and growth. In *Rethinking international trade*. The MIT Press.

Ohlin, B. (1924). *The Theory of Trade*. Stockholm: Handelns Teori.

Robinson, J. (1977). What are the questions? *Journal of Economic Literature, 15*, 1318–1339.

Schumpeter, J. A. (1934). The Fundamental Phenomenon of Economic Development. *Schumpeter, The Theory of Economic Development*, pp. 57–94.

Schumpeter, J. A. (1942). *Capitalism, Socialism and Democracy*, New York: Harper.

Smith, V. L. (1982). Microeconomic systems as an experimental science. *The American Economic Review, 72*, 923–955.

Ullberg, E. (2009). From personal to impersonal exchange in ideas – Experimental study of trade in organized markets for patents. *KTH TRITA-TEC-PHD 09-006*, 180.

Ullberg, E. (2010a). Du passage d'un système d'échanges "personnels" à un système d'échanges "impersonnels" des idées: une étude expérimentale (Encadré 3). In D. Guellec, T. Madiès, and J.-C. Prager (Eds.), *Les marchés des brevets dans l'économie de la connaissance.* Report for the Economic Counsel of the Prime Minister, Paris.

Ullberg, E. (2010b). From personal to impersonal exchange in ideas: An experimental study of patent markets with transparent prices. CESIS Working Paper Series #230.

Ullberg, E. (2010c). The problem of trading patents in organized markets: A dynamic experimental microeconomic system model and informal price theory. CESIS Working Paper Series #229.

Chapter 5
Conclusions

5.1 Introduction

In this chapter, I propose to summarize the theme of impersonal exchange in technology based on linear contact on patents and proposed informal theory for prices on such exchange as discussed in Chaps. 1 and 2. Key experimental results, presented in Chaps. 3 and 4, are then discussed with respect to the performance and behavioral properties of institutions and the patent system environment investigated in the experiment. The conclusions are summarized under general themes and experimental results.

5.2 General Themes

5.2.1 Investigating Design Institutions Toward Impersonal Exchange in Ideas

The theme of the research has been the transition from personal to impersonal exchange in ideas. Through the patent system, which was first developed in 1474, (technical) ideas became tradable in their own rights. This was an intentional economic policy to create incentives for importing (trading) technology to the then city-states and making the previously (in guilds) privately held knowledge public at the same time granting temporal private excluding, tradable and licensable rights on the published technology. Since then, the patent system has gradually developed toward stronger and more nationally and internationally tradable property rights, resulting in a change in economic organization with more and more specialized agents for invention, intermediation (trading), and innovation. As a result, the *coordination* (cooperation) of inventions and innovation activity is shifting from within a hierarchy to between hierarchies through *market with prices*. This market of

E. Ullberg, *Trade in Ideas: Performance and Behavioral Properties of Markets in Patents*, Innovation, Technology, and Knowledge Management 13,
DOI 10.1007/978-1-4614-1272-4_5, © Springer Science+Business Media, LLC 2012

interest is thus a *producer* market for patented technology. The experimental analysis of *dynamic* gains from trade in technology between specialized agents based on contracts on *patents* adds to the traditional static analysis on product monopolies organized in a hierarchy.

5.2.2 How Risk is Shared in the Contract as Coordination Takes Place in a Market with Prices

The main goal was to investigate the impersonal exchange of *technology* itself, given a patent right, studying how *people* divide up the values of a patent (the blocking, investment, and trade values) in the fixed fee and the royalty price in the (linear) contract used to trade the patents in the experiment and commonly used in real-world transactions.

The three patent values represent different ways of using a patent and differ in nature with regard to risk and uncertainty. Understanding these *risk-bearing* and *risk-sharing* preferences can hopefully lead to better integration of the information (messages) and rules (giving incentives) of institutions for higher social gains from exchange of technical ideas through the patent system. Vernon Smith summarized human social behavior, writing that "… exchange promises gains that humans seek relentlessly in all social interactions." (2004, p. 67).

5.2.3 Mechanism Design: The Integration of Information and Rules

The *proper integration* of information and rules in institutions are currently beyond game theory's reach. See Hurwicz (1973, p. 27). However, experiments can provide information on these outcomes on a case-by-case basis. Here, the approach to the mechanism design issue has been to systematically increase demand-side bidding – expressed as varied language (messages) – using institutions that are commonly known in similar one-dimensional versions. The results may therefore indicate which *information needs* and *institutional rules* for exchange contracts on patents are socially preferred, such as pareto optimality (P.O.). In this case, the microeconomic system is based on a dynamic concept, and "socially preferable" means that technology is *invested* in (not blocked), thereby maximizing the use of technology in society.

5.2.4 A Dynamic Economic System

The investigation is done using a *dynamic* microeconomic system design that integrates an *economic* environment based on the linear contract on patents and trading agents, an *institutional* environment of a primary and secondary impersonal market designs and a *legal* environment based on patent system principles (exclude, trade, multiple fields of use (claims), public disclosure of technology fields (disclosure), time limit and exclusive/non-exclusive license). The design is flexible in terms of mechanism designs and can be used for many experiments of similar nature.

5.3 Experimental Results

The results of the experiments, basic theoretical work and economic history, patent system review, and literature review are now summarized as propositions and conclusions:

5.3.1 Do Experiments Make Sense in a Complex Dynamic Environment Like the Patent System?

The first question answered was whether experimental economics could in fact be used to study trade in patents.[1] The results show that successful experiments can be designed and implemented for trade with patent rights. Subjects actually do trade these rights and markets clear, and coordination between specialized agents takes place in a *dynamic* experimental economic environment. This is promising since data on personal exchange licensing contracts are hard to obtain. The experiments therefore provide important information for drawing conclusions on mechanism design and contract issues.

Voluntary trade appears to need capital and bankruptcy rules to restrain participants from overbidding.[2] Strong property rights on the contracts traded was not enough. This result is consistent with prospect theory and "loss aversion" (Kahneman and Tversky 1979).

5.3.2 Why Are There No Organized Patent Markets Yet?

As noted, there has been a great deal of activity in this direction during the last 25 years, but as yet there is no organized market resembling anything near the stock

[1] Some experiments in relatively simple and static environments have been designed, but to my knowledge not in a dynamic system. Experiments include "patent races" and "spillover" effects, i.e., pre- and postpatent time. This study was entirely dedicated to during patent time.

[2] This problem, I have seen, is often solved by simply forbidding subjects to bid above their value. This is not the road taken here. Such rules are "impossible" to put in place in the field. In fact, they typically do not exist in the field. The idea has been to put in place rules that create *incentives* that in turn create efficient markets in this dynamic environment.

market. Buyers and sellers are uncertain whether they will be better off with public prices.[3] The results propose an answer to this question: it is very important which kind of *contract* is used, what messages and rules are used, and how the messages and rules are integrated so that the *two* dimensions in a patent value can be priced independently (blocking and investment values). The experimental results increased the dynamic market efficiency by about 100%, compared to the institution closest to "personal exchange," when demand-side bidding was introduced for the linear contract. However, before it will be possible to provide any helpful advice about what to do in the field, additional experiments are needed to continue to increase market efficiency.

5.3.3 What Are the Main Theoretical and Experimental Results?

The results are listed as propositions under contract/price theory, mechanism design, Trader dynamics. A separate section includes tentative policy proposals.

5.3.3.1 Contract–Price Theory

Proposition 1 A fix fee is paid for a patent (in the linear contract) that equals the blocking value of that patent.

In the institution with two-dimensional bidding (DA1), where the theoretical prediction is that the blocking and investment values aught to be independently expressed, the average price paid is 0.5–10% away from the second highest blocking value (the game theoretical prediction when using discrete and stacked induced values for bidders). The null hypothesis (fix fee equals the second highest blocking value) cannot be rejected under standard statistical criteria. Such a result supports the informal prices theory. Therefore, a key result is that a fix price could be calculated based on this "insurance value" against product competition based on the technology that is blocked.

For boundary conditions with high blocking, low investment values, and vice versa, the fix fee is less than the blocking value for high Block/low Invest and higher for low Block/high Invest. The fix is never bid to zero and the royalty component is rarely bid to zero.

Conclusion 1: Given proposition 1, the risk-bearing is shifted in the microeconomic system from the Inventor to the Innovators, i.e., ultimately to the consumers. This means that the risk of inventive activity decreases. The inventive activities would then compete for capital with innovative activities. With well-functioning capital markets (efficient allocation of funds given the risk in the investment), capital should, then reallocate in favor of inventive activities, resulting in more technology[4] and a more competitive technology market, producing more economically useful

[3] Personal communication with Prof. Steve Rassenti.

[4] The single most important factor for inventiveness in firms is the size of their research budgets.

technology. Such increase in useful technology and the public knowledge thereof will certainly create social gains.

Capital would equally flow to fund intermediaries, taking on the intertemporal risk between investment and invention (through the fix fee). Since technology is widely considered the single most important factor for economic growth,[5] impersonal exchange in patents may therefore benefit economic development by reallocating the risks in the economy *in favor of* the development of new technology.

This exchange mechanism of technical ideas thus provides an explanation for the very successful integration of science and technology in industrialized countries. Organized markets may therefore make the patent system more incentive compatible with social objectives in some cases where high specialization takes place. This would be the case for industries with high interoperability requirements (such as standards or component-based industries).

I would here like to quote Joan Robinson from "What are the questions?"(1977) p. 1337:

> Here is the problem. The task of deciding how resources should be allocated is not fulfilled by the market but by the great corporations who are in charge of the finance for development.
>
> These questions involve the whole political and social system of the capitalist world.

My contention is that an impersonal market in ideas provides a solution to this problem of allocation of capital through another, socially preferable, risk allocation process. This process uses the patent system as a *trade system* that provides incentives[6] to maximize the *use* of new technology and development through competitive prices.

Conclusion 2: Proposition 1 strongly supports the *informal price theory*, which predicts that a blocking value, viewed as an "insurance" against the loss of short-term market access, would be expressed in full in the fixed fee (not in the royalty). I therefore conclude that the patent's blocking value can formally be viewed as being similar to an insurance policy. This result is in line with the view that risk and uncertainty can be separated by "insurable risks" for which there is an insurance market (Knight 1921; Barzel 1986; Schumpeter 1994; LeRoy and Singell 1987), and formally similar to insurance theory (Arrow 1965; Smith 1968).

Both the risk and uncertainty are given a *price* in this linear contract market. One risk is based on "objective" probability, while the other is based on a more "subjective" probability (e.g., Friedman's 1976) linked to the capabilities of the specialized hierarchy to *manages* the investment risk. Perhaps managers, through people (knowledge), systems, and intellectual property can do a better job than a lottery.[7]

---

[5] See, for example, White, Lenski, Tofler.

[6] Incentives to invent to use (the standard analysis based on the hierarchical model), invent to sell to use (the licensing approach with two agents), and invent to *competitively trade* to use, which provides for dynamic gains from trade in technology (the dynamic licensing approach including trader agents, the focus of this study).

[7] However, Prof. Makridakis at INSEAD once suggested that flipping a coin is as good as hiring a manager to make strategic decisions. This was also Knights view on uncertainty.

5.3.3.2 Mechanism Design

Results for mechanism designs are summarized for linear prices, dynamic gains, and coordination through prices.

Three mechanisms were tested, differing according to the messages that could be sent by the demand-side and the bidding rules in the market. These design mechanisms resulted in different price outcomes (especially for high presumed patent validity). The contention here is then that two-dimensional bidding is required for an efficient price system. The ranking DA1 > DA2 > PO[8] holds generally for propensity for dynamic gains.

In a "robustness test" for DA1, the induced values were randomly varied (similar to two different technologies) and the price outcome clearly followed the fundamental values.

In the DA2, the higher fixed fee induced a behavior of higher propensity to invest. The agents therefore traded off some of the investment risk for a higher fixed fee to win the contract. Here, the risk transfer was higher than the blocking value, meaning that entrepreneurial risk was taken on by the Innovators. The institution then induces "compulsory investment" of the technology. This appears to be inconsistent with an efficient pricing system where the price must be the same for whatever use to which it is applied. In the robustness test, DA2 appeared to distinguish the different "technologies" somewhat correctly, though not as well as DA1.

In PO, a lower fixed fee also lead to a lower corresponding royalty. In this case, which is the most similar to today's personal exchange, there is not the same risk transfer as in the other institutions. One reason for the lower fix may be that subjects do not know (learn) how to negotiate a higher fix.[9] In the robustness test of PO, there was no difference at all. This means that high-value technology is priced at low-value technology level. The ranking is still maintained DA1 > DA2 >> PO.

The conclusion to this *dynamic* pricing behavior is that demand-side bidding would increase the accuracy in "discerning" between high-value and low-value technology.

Thus, the issue of robustness is not only a matter of convergence but also a matter of incentives, preferences, and subject behavior; what people can possibly learn about the other side to arrive at an incentive-compatible outcome. This question is at the heart of the challenges to creating efficient markets for patents, since patented technologies by definition are new, unique, and typically with many applications, therefore having different values.

---

[8] DA1 – "double auction" in both the fixed fee and royalty price (" two-dimensional bidding"), DA2, reservation value on fixed fee and bidding in royalty, PO, "posted offer" by seller and accept/reject by buyer.

[9] This proposition was made by Prof. Rassenti.

Coordination of allocation of resources for invention (production of technical information) through the linear prices also differs by mechanism and validity. The main result indicates that the "willingness to search" is proportional to demand-side bidding. The market clearly provides as a price signal in the experiments.

An interesting observation was made concerning social exchange. Most subjects stayed in the same technology areas that others or they have found (a kind of "loss aversion"), whereas some tried new areas. Their motivation (when asked) was "I always want to try new things" and "If I try something now, others may learn and then when it is my turn (in the game) I will also benefit." This behavior suggests, however weakly, that there is some social exchange taking place through the market and the information on which technology areas are valuable and which are not. Moving from personal to impersonal exchange apparently does not change peoples' desires to seek gains from social interactions (through patent disclosures and prices).

5.3.3.3 The Trader and Some Trade Dynamics

The Traders are able to outbid the Innovators using the fixed fee, not the royalty. This result is statically significant. They then are able to resell the split contracts to the Innovators they just outbid. The Inventors, from whom they buy the contract in competition with the Innovators, prefer a higher fix feed fee to a royalty. This is clearly seen in the results (from DA2), a result possibly consistent with prospect theory and "loss aversion," where insurance (fixed fee) is preferred to gambling (royalty). The Innovators have the opposite preference, preferring royalty to fixed fee (from DA1). The Trader can match both these preferences through splitting the contract into different fields of use, geographic markets, etc, taking the inter-temporal risk between fixed fee now and royalty later. The end result, with rational agents with some loss aversion, is that the risks are shared in more consumer markets, i.e., a broader market access is achieved for the technology. The Trader role is therefore incentive compatible with the social intentions of the patent system.

5.4 Further Research Propositions

This research has aimed at showing that economic studies could be meaningful using experiments, gathering experimental evidence on prices, dynamic gains from trade and coordination through markets, and looking into the complexity of trading patents in dynamic environments with impersonal exchange mechanisms.

Building on the results and conclusions of the research performed there are still many open questions in the field of trading patents. To take the price theory, market designs and the historic understanding of the patent system as a trade system further, a list of future research is here been proposed. See Table 5.1.

Table 5.1 List of future research projects

1.	Further develop the *trade history* of the patent system using an exchange analysis
2.	Further develop the *mechanism design* for new experiments
3.	Further develop the understanding of the contract, focusing on the *incompleteness of the contract* traded. This would require trust between firms
4.	*Explore the tension* between blocking and investment bidders
5.	Develop a *formalized model* of the dynamic experimental design
6.	*Bidding adjustment process*: Explore bidding strategies in linear auction markets
7.	Develop a *formal price theory* for the linear contracts on patents
8.	Introduce competing technologies on the *sell side*
9.	Investigate an alternative to the static patent time system. Experiments with *varying patent time* may reveal incentives and (dynamic) values to the economy. This was in fact the practice when the patent system began

These studies demand an interdisciplinary approach of economic and legal experiments.

References

Arrow, K. J. (1965). *Aspects of the theory of risk-bearing*. Helsinki: Yrjö Jahnssonin Säätiö.

Barzel, Y. (1986). Knight's "Moral Hazard" theory of organization. *Economic Inquiry, 25*, 119–120.

Friedman, M. (1976). *Price theory*. New York: Aldine de Gruyter.

Hurwicz, L. (1973). The design of mechanisms for resource allocation. *The American Economic Review, 63*, 1–30.

Kahneman, D., & Tversky, A. (1979). Prospect theory: An analysis of decision under risk. *Econometrica, 47*, 263–292.

Knight, F. H. (1921). *Risk, uncertainty and profit*. New York: AM Kelley.

Leroy, S. F., & Singell, L. D., Jr. (1987). Knight on risk and uncertainty. *The Journal of Political Economy, 95*, 394–406.

Robinson, J. (1977). What are the questions? *Journal of Economic Literature, 15*, 1318–1339.

Schumpeter, J. A. (1994). *History of economic analysis*. London: Routledge.

Smith, V. L. (1968). Optimal insurance coverage. *The Journal of Political Economy, 76*, 68–77.

Smith, V. L. (2004). Human nature: An economic perspective. *Daedalus, 133*, 67–76.

Chapter 6
Policy Propositions

6.1 Policy Propositions

The historic narrative in Chap. 1 proposed a reexamination/reevaluation of the motivating factor behind the patent system to what appears to have been an economic policy to import (trade) technology to enhance economic development, eliminating the granting of manufacturing privileges, which created product monopolies, and as a "guild buster." The hope was to create a more efficient system for the development of and public access to (new) technology, by means of a competitive market in technology itself (not product monopolies/privileges) granting an exclusive, transferable, and licensable right in exchange for publication of the same technical solution to a problem. The hope was that this both social and market-based exchange system would be more efficient (and cheaper for the then city state) than the previous government subsidies and secrecy. The system introduced *the principle of a social and market exchange* for technology based on private property rights to new and productivity enhancing inventions, in a similar way as to land and capital. The declaration of an excluding, tradable, and licensable right on economically useful technology for disclosing the invention to society (and thus making the previously secretly held technology public knowledge) reduced the risk for the inventor to exchange commercially and socially the technical idea (previously held privately as a trade secret), providing an institution with incentives to invest in research and technology and to grow the stock of knowledge and technology in society, the goal of the economic policy.

This may have come in response to financial pressure on the state who wanted a less costly policy for technology development than previous government subsidies and secrecy (which was defended with draconian measures at the time) (Kaufer 1989). The new policy reduced the cost both in the form of less government grants by the state, and for the inventor the cost of keeping the knowledge secret. The exclusive right granted for a limited time, created an economic incentive for inventors to bring their inventions to public knowledge, incentivizing more inventions, and producing with the higher productivity beneficially to the economy of the state.

E. Ullberg, *Trade in Ideas: Performance and Behavioral Properties of Markets in Patents*, Innovation, Technology, and Knowledge Management 13,
DOI 10.1007/978-1-4614-1272-4_6,

Today, this is still a policy issue and an issue of discussion, using markets in technology to develop and share technology or government grants and secrecy. The hope is that this book has shed some lights on the gains from the exchange in technology, socially and economically, as a trade system, and additional knowledge about the performance and behavioral properties of these institutions.

The economic performance and behavioral properties have been studied in a 3×2 design, with three design market institutions and two levels of patent strength (presumed validity). This chapter addresses policy issues related to the performance and behavioral properties of such markets.

The main experimental results support the proposed theory that the blocking value of a patent ("sitting on" it) if formally similar to insurance, which raises implications for the right to perform such auctions, limits of contingency contracts, levels of presumed validity of patents, and intermediary trader capital demand. The main result with respect to validity (strength) is that the propensity to invest decreases drastically with low validity (blocking dominates) and prices then tend to be more similar between institutions (as assumed by neo-classical theory). Also what is paid in royalty goes down (there is a risk of loss). However, with stronger patents, differences in outcomes between the institution are statistically significant, with much higher propensity to invest, split, and exchange, in the technology, as bidders can express the dual patent values, suggesting that both the social and economic gains from the patent system are sensitive to the information and mechanisms/rules used in organized exchange (contrary to neo-classical theory). In these outcomes, subjects have a higher propensity to invest rather than block, and are thus more incentive compatible with social gain measures, such as optimal use of the technology in products. These and results presented in Chap. 5 lead to some tentative policy proposals. The propositions here are both general in nature drawn from the results and specifically to the institutions and patent validity tested. The propositions contribute to the policy question of how to coordinate what the patent system does with the purpose of the patent system.

The first propositions concern the continued specialization in the management of financial and entrepreneurial risk in the economy. The patent system appears to change the economic organization by allowing specialized agents to trade ideas through personal and impersonal exchange and, thereby, how the risks are allocated in the economy. There are two issues here: the *contract* used and the introduction of (intermediary) *Traders*.

The first issue brought forth in this research is that the prices, using a linear contract to trade patents, provide for allocation of risk-bearing (fixed fee) and risk-sharing (royalty). The outcome of the allocation of risk-bearing was that the inventor was paid for the blocking value (or something close to that), reducing the risk in the production of information. There is, therefore, an incentive created to take more risk in production of information, as capital would flow toward this activity central to the economy.

The second factor of risk-bearing is the introduction of Traders. This shift has a *systemic* effect as Traders assume intertemporal risks – between the invention and the innovation. They are in effect financing the trade in patents. The risk lies in the fix fee of the linear contract that has to be covered by the fix fee or royalty from the Innovator, for which there may be a difference in time (market liquidity and new products launches).

There are two immediate propositions from this: (a) the use of capital in trading (intermediating) to enhance multiple use of the technology and increase the propensity for dynamic gains, bridging the time gap between inventing and innovating, and (b) a new form of legal entity with incentives to innovate with through lower marginal cost of capital.

Proposition 1 *Finance the trade in ideas*[1] which amounts to a "new" use of capital. Since the contracts are contingency contracts, it is reasonable that there is an appropriate capital cushion held by the Trader to cover for the level of risk borne. The gains from this capital allocation would come from higher use of technology, thus an incentive compatible capital allocation. A secondary effect of the higher use of technology is that the cost of capital goes down in the economic system. The *systemic* risk is reduced (in a CAPM/ICAPM sense[2]). This entity is thus the subject to a new classification and capital requirements.

Proposition 2 *The inventive company* whose risk is lowered also depends on the willingness to invent. The results indicate that this is a matter of institutional design but it is also likely a matter of willingness to take on risk. The proposal here is to create a *new* type of legal entity, the creative company, which is *exempted* from *any* tax (on profits) but for this privilege, is asked to have a (much) higher capitalization than the traditionally hierarchically integrated firm.[3] The tax is instead taken out at the consumer end (VAT) or intermediary end (Innovator) of the system where the risks are lower. This would give incentives to take on and mange more inventive risks in the system and would be in favor of economic development based on technology. Such an analysis is similar to (Bernoulli's 1738; 1954) thoughts on the "new theory on the measure of risk." Incentives to increase the willingness to search further for a yet better technology focus may then be less restrained by loss aversion (Kahneman and Tversky 1979), as the only way to get a return (without paying taxes) is to invent. This proposition is subject for further investigation for suitable level of capitalization given the higher risk in inventing.

The next propositions address the structure of the information in the patent system as a key for trading.

---

[1] This question is elaborated in a previous chapter.

[2] Capital asset pricing model and intertemporal CAPM which makes a distinction between a systemic financial "risk free" rate ("alpha") and individual risk ("beta") based on share price volatility. The intertemporal version allows for other "betas" than the financial ones. See Merton (1973) and Fama (1996). A potential use of CAPM with respect to patents is briefly discussed in: Ullberg (2004).

[3] Similar ideas are being tried now by France, who recently (2007) introduced a tax break for 50% of investments in research. However, this is a system with the current corporate organization.

Proposition 3 *Inclusive development through international standards for presumed patent validity.* In order to continue development of international trade in patents, similar valuations of the same rights would be preferable. Standards for evaluation and contracts would reduce risk in prices and provide a better proxy for values.

Such an agenda is a potential *economic development* agenda, involving developing nations. The approach would represent an alternative route for market access to the global economy for such countries.

Proposition 4 *Claims structure suitable for "splitting" and "bundling."* Maybe the most important aspect in relation to tradability is the claims *structure*. It is conceivable that a structure more targeted to "separable" fields-of-use would enhance the clarity in what rights contain and what rights are easily licensable. This would be something like a "bundle" granted of rights. To explore this bundle in legal terms would in my view have direct impact on the gains from trade coming from the patent system.

Proposition 5 *Disclosures with cross references.* The disclosures are the basis for social exchange of what is invented and a key issue for trade is therefore that the information regarding what is invented is clearly written. A certain technology is almost always patented in several patents, one reason being to avoid disclosing the technology as a whole. This was the opposition of the guilds to the patent system. To create incentives to disclose, references to the other parts of the technology could be made in the disclosure. Perhaps there is a new or modified technology structure code needed that focuses on the anticipated use of the technology. Such disclosure would reduce the risk for the buyer that "all" technology needed to produce is revealed by the seller.

These three and other issues related to the structure of information and its legal validity could be suitable for a patent *trade* policy.

Proposition 6 *Disclosure of ownership of patents.* There is currently no updated information on who owns what patents world-wide. Standards for recording changes in ownership vary. Knowing who owns certain technology is central to understanding risks of infringements, etc. Creating a incentive to update ownership of patents would be a first step to facilitate buyers and sellers to find each other and to evaluate each others holdings.

The next two propositions are concerned with international exchange and economic structure.

Proposition 7 *The technology trade policy applicable to developing countries.* The patent system gives today market access for new technology to most countries in the world. The system is however resisted to some extent by many of the developing countries as allegedly being used (or targeted) to protect Northern "export" of, among other things, reverse engineered products (especially in biotech) to the countries of "origin" of the knowledge, i.e., the South. Looking at the trade history of the system and principles of the system, it provides the mechanism for *any* nation to get global market access for their ideas, creating a leverage of national human capital, in particular on technology. The risk shifting indicated by the experimental results provides a cash-flow for the inventing countries and a focus on the "production of information" for economically useful knowledge that *any* country can participate in.

A national patent system together with efficient market mechanisms may be able to tip the balance in favor of many countries with comparative advantage in certain knowledge to attract foreign direct investments (FDI) in research today. Such a development agenda needs an international patent system that is equitable for developing nations, i.e., a *trade-based* system. Developing this *trade principle* may therefore provide an alternative way to market access for developing countries.

Proposition 8 *Could economist have gotten it wrong?* In the most recent major overhaul of the patent system in 1952[4] (USA), the balance in the patent system was shifted with respect to the criteria for invention from "flash of genius" to "nonobvious" ("inventive step" in Europe). A judge commented then that "long toil now had replaced flash of genius" (Kingston 2004). This can be seen a *political economic* decision where companies at the time were largely integrated corporations, leaning heavily on what appears to be the hierarchical model of invention and innovation developed during the twentieth century. This model had now "outpaced" the inventor model observed by Lamoreau (1999) to have created a market in technology in the nineteenth century.

The decision appears to have moved the system closer to a *product* monopoly, something that was abolished in the UK in 1623, statues of monopolies paragraph 6 (and apparently not intended in Venice in 1474). Reducing the "inventive step" also allowed "marginal inventions" to be protected, thus approaching a protection for product *development.*

Perhaps, and here is the point, the *economic analysis* focused on (product) monopoly and divergence between social and private gains had failed to take into account the *dynamic* effects of the patent system, as is argued here, and thus such a policy change may have been misinformed by possibly understating the dynamic effects of trade?

Today the industry structure is reshaping again in favor of a more integrated cooperation of *independent* specialized firms with respect to invention and innovation (since 1970s). The need for interoperability between specialized actors

[4] As this book is finalized (2011), the US House of Representatives have passed the greatest change since 1836, abandoning the principle of "first to invent" for "first to file" for priority date for an invention, effectively moving the patents system to the European (and rest-of-the-world) standard. It is thus a step in global harmonization of the patent law, something tried prior to 1883 but instead broad principles became the basis for the Paris Convention. It does address some questions of patent quality through operational questions such as patent processing, post-grant review and limiting frivolous lawsuits on alleged patent infringements. These measures will probably impact presumption of validity though. A key aspect is thus to reduce uncertainty for patent holders, something here argued to be essential for economic development based on technology. This new law has resemblances with the strategy chosen by EPO in 2002 to reduce risk for the users of the patent system in order to boost economic development. See www.USPTO.gov for more information on the new US law.

heavily depends on patents to coordinate their activities and products and services in most industries already today. An alternative policy would thus be to develop more internationally tradable patents, allowing for exchange of patents to take place at a lower risk. If these experimental results can be repeated, and expanded, maybe there are reasons to look at this change again?

6.2 Future Research to Clarify Further Policy Issues

The economic policy of import (exchange) in technology and honoring inventors that is here argued was the motivating factor for the patent system, challenges alternative policy solutions, to grow and effectively use technical knowledge in society and thereby sustaining economic development. In particular, government subsidies and secrecy are questioned as effective policy instruments when gains from specialization are taken into account from a technology exchange policy based on the patent system. It is my contention that the problem of social and market gains from trade in ideas requires a systemic view. Coordination between hierarchies through exchange and publication has to be compared with government subsidy and secrecy granted to selected hierarchies. The following suggested research topics are hopefully helpful in advancing our understanding of economic development through technology development, exchange, and use.

6.2.1 Role of Patent System in Development

An expanded economic historic narrative of the patent system and its role in economic development through time would provide information on the proposed economic policy thesis of the patent system as a trade system. A rigorous analysis of the key stages in the patent system history outlined in Chap. 1 would be expanded and the political economic discussion elaborated. Then, given availability of data, a regression based on the social and market mechanisms would shed light on exchange in ideas as a policy instrument.

Such an analysis would beg for questions regarding technology typically kept secret: military technology. Would a less restrictive policy be testable along the lines of an imagined Bayh-Dole Act for military technology no longer deemed strategic but potentially valuable for civil economic use? This is in effect what DARPA appears to do as they include inventive ideas from companies in its program goal to create new industries and sources of economic growth.[5] The purpose would be to investi-

[5] See the DARPA Web site: Published minutes from PCAST meeting at NAS on January 7, 2010, Item 4.

gate policies which could allow disclosure and thus patentability of some "spin-off" or for strategic purposes "outdated" but for civil purposes productive technologies. Prior to the patent system, high wages and social benefits were offered to attract technical competence to military production (Kaufer 1989, pp. 4–5). Now, coordination on formerly strategic technology could be done through the patent system.

6.2.2 Insurance

A key finding from the experimental study is that the proposed price theory of a patent's blocking value as an *insurance* contract is supported by the experimental data. In the broader scope, technology is "insurance" for market access. Formalizing the insurance price theory could therefore be useful to generate additional testable hypotheses to complement findings. Insurance theory has broad foundations that could be combined with the intuitive narrative presented here (Smith 1968; Arrow 1971).

6.2.3 Trust Between Firms

Binding and complete contracts are typically traded in experimental studies. This is impossible in real-world transactions. Contracts are incomplete due to future contingencies not known at the time of contracting and can be challenged in courts. This is especially important with respect to contracts on patents, where the previously granted patents claims can be challenged in court and amended and future usage of intellectual property is difficult to specify. Trust in each others actions not specified is necessary to overcome this problem. To address this concern, a study of incomplete contracts could shed light on the information and mechanisms' firms use to overcome the incompleteness and put trust in each others actions. This problem has extensively been studied for personal trust between individuals but a little trust between firms, a sociological constructs, and a more impersonal trust. Understanding these mechanisms better would inform policy on exchange in technology to reduce risk (and cost) in transactions in technology based on the patent system, thus creating a more performing economic system. This question may very well be at the heart of economic development based on sustained technology (knowledge) development in a global, more impersonal, economic system developing today.

6.3 Concluding Remarks on Experimental Research

Seeing the patent system as a trade system is perhaps to some extent a novel approach to economic science but not to its more recent de facto use or legal and economic history. Perhaps one could say that the political economic history tells a compelling story of trade and gains from trade in technical ideas yielding social and market gains. To expand this complex system design, experiments may be the only way currently available to forward the understanding. Theorists are beginning to use this method of examination of new theories where data are scarce or unavailable.[6] As markets develop we can learn from them; until then, however, the laboratory seems to provide basic and principally true information useful for developing more explicit theories of the trade aspects of the patent system in its own rights.

References

Arrow, K. J. (1971). *Insurance*. Essays in the Theory of Risk-Bearing: Risk and resource allocation.
Bernoulli, D. (1738 (1954)). Exposition of a new theory on the measurement of risk. *Econometrica, 22*, 23–36.
Fama, E. F. (1996). Discounting under uncertainty. *The Journal of Business, 69*, 415–428.
Fehr, E., Hart, O., & Zehnder, C. (2009). Contracts, reference points, and competition-behavioral effects of the fundamental transformation. *Journal of the European Economic Association, 7*(2–3), 561–572.
Kahneman, D., & Tversky, A. (1979). Prospect theory: An analysis of decision under risk. *Econometrica, 47*, 263–292.
Kaufer, E. (1989). *The economics of the patent system*. Switzerland: Harwood Academic Publishers.
Kingston, W. (2004). Why harmonisation is a Trojan horse. *European Intellectual Property Review, 26*(10), 447–460.
Lamoreaux, N. R., & Sokoloff, K. L. (1999). Inventive activity and the market for technology in the United States, 1840–1920. NBER Working Paper.
Merton, R. C. (1973). An intertemporal capital asset pricing model. *Econometrica, 41*, 867–887.
Smith, V. L. (1968). Optimal insurance coverage. *The Journal of Political Economy, 76*, 68–77.
Ullberg, E. (2004). Is there a patent beta? Cost of capital evaluation of patent active and patent non-active companies. Working Paper Note.

[6] For a recent example of this, see Fehr et al. (2009).

Epilogue

A summary of this book was included in a report on "Patent Markets in the Knowledge Economy" for the Economic Counsel of the French Prime Minster (Ullberg 2010b) and is reprinted here. The notion of the creative company is further discussed as are implications for future North-South trade in ideas with the purpose of a technology transfer mechanism to the poor.

Recent bidding on the Nortel patent portfolio (June 2011) following the bankruptcy of Nortel Networks Corp., the Canadian telecom equipment manufacturer, illustrates the dynamics of the market in ideas investigated in the experiment. See referenced article from Bloomberg News and others below. All the basic feature of the experimental market is illustrated here: bidding, blocking, investing (trading), and innovating for market access. Interestingly, a fixed price is asked for the portfolio (probably set by the bankruptcy rules). That would then correspond to a mix of the blocking value and the investment value of the technology. These bankruptcy auctions may benefit from a combinatorial auction model mentioned in this book as a possible next step in patent market analysis using experiments. Buyers would be able to create valuable *bundles* of patents and bid on them. The *mechanism* would maximize the value of the auction as a whole, solving allocation problems "impossible" to solve in a sequential manner (by reselling). Such auctions hold promises of increased revenues for the state administrating the bankruptcy. Such mechanisms and others, hold promises for future Inventors, Intermediaries, and Innovators operating in the patent markets.

A second research project is planned to address the question of incompleteness of contracts traded which will result in a follow-up experiment on this topic. In most experiments and theory, contracts are binding and complete. This is far from real-world situation though where it is impossible to formulate a contract that includes all contingencies and possible future unknown events. Firm still trade ideas though, along the lines explored in this study, showing that there are other mechanisms involved, creating trust in each others actions. This trust between firms will be the topic of this second study.

The growth of technology is at the heart of the problems of political and economic performance today. This book has used experimental economics to advance

E. Ullberg, *Trade in Ideas: Performance and Behavioral Properties of Markets in Patents*, Innovation, Technology, and Knowledge Management 13,
DOI 10.1007/978-1-4614-1272-4,

institutional learning of the behavioral properties of the complex and dynamic trade in ideas based on the principles and practices of the patent system. Two initial experiments have been conducted and much remains to be done. The attempt has been to make a positive economic analysis useful to inform policy, including North-South trade, leveraging human capital formation through exchange in technology.

From Personal to Impersonal Exchange in Ideas: An experimental Study[1]

Du passage d'un système d'échanges «personnels» à un système d'échanges «impersonnels» des idées: une étude expérimentale.

Par Eskil Ullberg, Visiting Senior Research Scholar, Interdisciplinary Center for Experimental Science, George Mason University (USA).

L'objet de ce travail est d'étudier dans le cadre d'un protocole expérimental le passage d'une logique d'échange des idées fondée sur une relation «personnelle» à des relations «impersonnelles.» Autrement dit, la question posée est celle de l'émergence d'un marché organisé des idées fondé sur un système de prix et régi par un système de protection des idées utilisant les brevets. Le passage d'un système de transactions s'inscrivant dans une «hiérarchie» (en utilisant la terminologie de Williamson) à un système de coordination par les prix n'est souhaitable dans le cas des transactions sur brevets que si cela permet de réduire le risque pour le système économique dans son ensemble et si cela procure des gains dynamiques (en élargissant le champ d'application des nouvelles technologies à un plus grand nombre d'utilisateurs et en allouant les technologies aux utilisateurs les plus efficaces). Une approche utilisant l'économie expérimentale est utilisée afin de définir les conditions pour que de tels échanges puissent exister et qu'ils donnent lieu à des gains dynamiques. Ce travail représente un des premiers essais développés en économie expérimentale portant sur des échanges organisés de brevets. Le modèle expérimental, la théorie des prix sous-jacente ainsi que le logiciel utilisé dans les 40 sessions (utilisant des incitations financières et faisant appel à des sujets de ICES-George Mason University) correspondant à deux expériences différentes sont décrits dans ma thèse de doctorat («From Peronal to Impersonal Exchanges in Ideas – Experimental Study of Trade in Organized Markets for Patents,» KTH, 2009).

Ce travail de recherche fournit un certain nombre de résultats concernant les effets de tels marchés dans le cadre d'un «design» expérimental qui explicite précisément la nature du contrat sur lequel porte l'échange ainsi que les règles («mechanisms designs») de fonctionnement de ces marchés[2]:

---

[1] Reprint from: Ullberg (2010b).

[2] On soulignera que la littérature expérimentale montre que le prix de marché converge vers un équilibre concurrentiel pour un nombre de participants beaucoup plus faible que ne le suppose la théorie de l'ordre de 3 à 6.

1. Il apparaît tout d'abord que si la demande d'enchère portant sur les contrats de brevets est concurrentielle alors cela conduit à un transfert de risques des inventeurs vers ceux qui utilisent ces inventions (les innovateurs) et par voie de conséquence vers les consommateurs. Les activités d'inventions sont donc plus profitables (car le coût du capital est plus faible). L'utilization des nouvelles technologies produites augmente aussi car les risques sont répartis sur un nombre plus élevé de marchés de biens finals et donc sur un nombre plus élevés de consommateurs. Une conséquence que l'on peut attendre de tout ceci est une augmentation des dépenses de R&D et du nombre d'inventions ; ce qui en retour devrait augmenter le stock de technologies échangées sur ce type de marchés et augmenter la concurrence (ce qui a un effet positif sur la croissance dans ce type de configuration).
2. L'existence d'intermédiaires spécialisés dans les activités consistant à créer des contacts entre acheteurs et vendeurs, à acheter des licences de brevets, les «diviser» en licences spécialisées destinées à de multiples marchés de produits, et à revendre à leur tour ces licences à des entreprises qui veulent mettre en œuvre les inventions correspondantes sont des agents «critiques» pour créer des gains dynamiques dans l'économie car ces intermédiaires ont pour seule incitation de maximiser l'usage de la technologie (dans des domaines, sur des lieux et à des moments du temps différents). Les intermédiaires sont capables de faire cela parce que sur un marché où les transactions sont guidées par un système de prix leur présence permet de réduire le «portage» du risque en le «dispersant» sur de multiples utilisateurs ayant des usages différents et appartenant à des zones géographiques différentes. Quand les intermédiaires entrent sur le marché, l'efficience dynamique augmente de 100 % par rapport à un système de transactions «personnelles.» Les résultats expérimentaux montrent que les intermédiaires sont plus enclins à entrer sur un marché de transactions «impersonnelles» que sur un marché de transactions «personnelles» (en d'autres termes, l'activité d'intermédiation qui est source de gains tend à se développer plutôt sur des marchés où les transactions se font de façon «impersonnelles»).
3. Troisièmement, un contrat linéaire (qui se caractérise par une redevance fixe et une partie variable de type royalty) est nécessaire pour définir le prix d'un brevet. Le droit d'exclure confère en effet au brevet une double valeur: le droit pour celui qui le détient d'utiliser de façon exclusive la nouvelle technologie en développant de nouveaux produits de façon à se donner un avantage concurrentiel sur les concurrents (valeur d'investissement ou «investment value») ou au contraire de bloquer d'autres entreprises en les empêchant d'avoir accès à cette technologie («blocking value» ou valeur de blocage). Une théorie des prix est développée dans la thèse qui montre que la valeur de blocage peut être appréhendée d'un point de vue formel comme un contrat d'assurance et la valeur d'investissement comme une option stratégique. La valeur de blocage permet à l'entreprise de s'assurer à court-terme contre une perte irréversible de marché. La valeur d'investissement est plus proche d'un calcul fait en situation d'incertitude (de type VAN en avenir incertain). Le risque associé à la valeur d'investissement n'est pas transférable («non assurable») et doit donc être partagé (une façon de procéder à ce partage de risque est de se mettre d'accord sur un montant ou

pourcentage de royalties). Les résultats de l'expérience montrent que ce que paient les acheteurs sous la forme d'une redevance fixe correspond à la valeur de blocage alors que la valeur d'investissement correspond au montant (pourcentage) de royalties (dit d'une autre façon, le risque est «transféré» dans la partie fixe de la redevance alors qu'il est «réparti» dans la royalty du contrat linéaire). Ceci explique pourquoi en autres choses il y a un grand nombre de brevets non utilisés mais toujours valides (dans le sens où leurs propriétaires ont redemandé le renouvellement – jusqu'à 50 % des brevets). Les expériences montrent qu'un système de brevets qui permet de garantir la validité du brevet (ceci dépend évidemment de la qualité de l'organisme qui délivre les brevets comme des décisions prises par les courts de justice) est considéré comme crucial par les sujets participants au jeu expérimental pour utiliser «l'option d'investir» et donc pour l'efficacité dynamique du système dans son ensemble.

4. Quatrièmement, pourquoi n'observe-t-on pas aujourd'hui de marchés des brevets organisés où les transactions seraient faites de façon impersonnelles ? Une raison est que les acheteurs et les vendeurs ne sont pas surs de pouvoir tirer de gains quand les prix sont rendus publics. Un point très intéressant, cependant, est le rôle joué par les intermédiaires dont on a vu qu'ils étaient un élément essentiel pour que des gains dynamiques soient possibles: en effet, il s'est révélé difficile pour la plupart des sujets impliqués dans l'expérience de jouer ce rôle d'intermédiaire, ce qui suggère l'importance de l'apprentissage par l'expérience. Expérience qui pourrait être aussi développée sur des «marchés tests» et qui est nécessaire avant que des marchés organisés ne puissent procurer toutes leurs potentialités.
5. Cinquièmement, les expériences montrent de façon intéressante que l'on a absolument besoin de capital pour échanger sur des marchés «impersonnels.» Le passage à ce type de marchés fait que les intermédiaires de marchés (traders, brokers, etc.) vont supporter davantage de risque qu'avec des transactions personnelles. Le fait de doter les sujets avec un capital (en quantité limitée) et de rendre possible la faillite permet d'éviter que ces derniers ne fassent des offres trop élevées (conduisant à une destruction du marché). Ce qui est en accord avec la théorie prospective («prospect theory») développée par Kahneman et Tversky. Plus précisément les intermédiaires vont procéder à un transfert de risque dans le temps entre inventeurs et innovateurs. Ce risque doit un couvert par du capital de façon à réduire les comportements trop risqués ou trop spéculatifs.
6. Finalement, les expériences suggèrent que des systèmes nationaux concurrentiels sont préférables à une harmonization des réglementations nationales sur les brevets. Le pays dont le système de brevet est faible (du point de vue de sa capacité à rassurer les participants sur la validité des droits attachés aux brevets) est incité à devenir meilleur avec des systèmes de transactions sur brevets concurrentiels. Ceci permet d'augmenter le nombre de technologies échangeables. Le travail expérimental montre au contraire que l'harmonization des règles nationales peut conduire à ce que le pays fort (du point de vue de la validité) devienne plus faible, ce qui réduit le nombre de brevets échangés sur le marché (la mise en

place de normes communes plutôt qu'une harmonization complète des législations nationales permet de laisser la concurrence jouer).

Un certain nombre de propositions peuvent être faites sur la base de ces travaux. En premier lieu, les expériences ont montré toute la difficulté de définir des règles d'échanges standardisées compte tenu de la spécificité de chaque brevet (ces règles devraient dépendre en particulier du domaine technologique couvert par le brevet). On doit donc s'attendre à être confrontés au même type de problème dès lors qu'il s'agira de passer de l'expérience en laboratoire à la réalité. En second lieu l'importance des intermédiaires sur des marchés «impersonnels» conduit à introduire des réglementations concernant ces intermédiaires (en particulier en imposant un minimum de capital). En troisième lieu, un marché «impersonnel» des brevets suppose la reconnaissance par les participants à ce marché d'un outil d'évaluation des brevets (transparence) ainsi que la mise en place d'un système juridique garantissant la validité des brevets afin de sécuriser les transactions et leur permettre de se développer de façon plus transparentes. La définition d'un certain nombre de principes communs entre pays (plutôt qu'une harmonization des systèmes nationaux de protection par les «brevets») concernant les transactions sont souhaitables pour un bon fonctionnement du marché. Enfin, nous proposons la création d'un nouveau statut d'entreprises dont l'unique objectif serait de se consacrer à l'activité d'invention («creative company»). Ces entreprises seraient exonérées d'impôt sur le bénéfice des sociétés (pour réduire le coût du capital) mais dotées d'un capital social minimum qui serait relativement élevé (par rapport à des entreprises de production). Enfin ce travail expérimental met l'accent sur l'intérêt d'une meilleure protection de la propriété intellectuelle dans les pays du Sud ainsi que sur le développement d'échanges de brevets mutuellement profitables entre le Nord et le Sud comme soutien au développement (les échanges d'idées contribuant à la formation de capital humain et donc à la croissance économique).

Traduction: les auteurs (après relecture par Eskil Ullberg)

The Nortel Patent Auction

Some news on the Nortel Patent auction that illustrates what firms active in patenting and patent licensing do to access markets through technology.

Ericsson Said to Weigh Entering Contest for Nortel Patents
http://www.bloomberg.com/news/2011-05-17/ericsson-said-to-consider-entering-competition-for-nortel-patent-portfolio.html

In this article, major firms and intermediary traders bid on the patent portfolio of Nortel to get market access for their products, preventing others from blocking, possibility to split the portfolio and license some or simply license the patents as a business proposition. The three roles used in the experimental study can clearly be identified – inventing (Nortel) – investing (producing and nonproducing bidders) – innovating (producing firms). More news on this auction can be found at:

Google Seeks Nortel Patents in Defensive Move
http://online.wsj.com/article/SB10001424052748703806304576242722516909158.html?KEYWORDS=google

Patent Portfolio Auctions from Commerce One and Nortel Networks
http://info.articleonepartners.com/blog/bid/56700/Patent-Portfolio-Auctions-from-Commerce-One-and-Nortel-Networks

Nortel Network's Patent Portfolio Worth 900 million dollars or More
http://www.coatsandbennett.com/2011/04/nortel-network%E2%80%99s-patent-portfolio-worth-900-million-or-more/

Nortel Files Bankruptcy in Delaware and Seeks Similar Relief from Creditors in Canada
http://delawarebankruptcy.foxrothschild.com/2009/01/articles/bankruptcy-case-summary/nortel-files-bankruptcy-in-delaware-and-seeks-similar-relief-from-creditors-in-canada/

Nortel Sale Is Biggest in Booming Patent Market
http://www.marketwatch.com/story/nortel-sale-is-biggest-in-booming-patent-market-2011-06-27

Reference

Ullberg, E. (2010b). Du passage d'un système d'échanges "personnels" à un système d'échanges "impersonnels" des idées: une étude expérimentale (Encadré 3). In D. Guellec, T. Madiès, and J.-C. Prager (Eds.), *Les marchés des brevets dans l'économie de la connaissance.* Report for the Economic Counsel of the Prime Minister, Paris.

About the Author

Eskil Ullberg, PhD is a Visiting Senior Research Scholar at the Interdisciplinary Center of Economics Science (ICES) at George Mason University. He has previously worked as a management consultant with strategy for 20 years, specializing management of risk and the patent system, participating actively in strategic joint work with the European Patent Office for economic development in Europe. As consultant Eskil has worked for a wide range of firms, government agencies, and international organizations on topics from strategy to development, focusing on the strategic use of information technology and patents (intellectual property) in creating competitive advantage for firms and nations in the global idea economy. His research interests currently focus on technology exchange between North-South and the use of the patent system, leveraging human capital formation in today's global economy. He has a MSc in Physical Engineering from Univ. of Uppsala, Sweden, an MBA from INSEAD, France, and a PhD in Economics from the Royal Institute of Technology (Sweden) awarded based on the experimental research and studies at ICES, GMU, USA.

About the Author

Appendices

Appendix 1

Software

A special software was developed for use in the two pilot tests and the two experiments. The pilot tests were used to verify: (1) trading a linear contract; and (2) the trade role. These programs were developed in Excel. The two experiments were developed in Visual Basic 6.0.

Due to its complexity, and endogenous decision-making, which affects the actual flow of the game, a *state machine* solution was developed. In each state, certain conditions have to be met to move to the next state. In this way, the program runs consistently in each state and then moves to run in the next state. The decisions of the participants thus decide which state to move into next, and which piece of software to execute at each time. This simple but powerful principle solves the whole endogenous problem of the programming. The idea was suggested by ICES programmer Lance from California.

The *state diagram* used in Experiments 1 and 2 are reproduced below.

A *data model* was also developed for the data handling during the experiment and the collection. Special data files were used to store data.

The *case diagram* was used to document the complex actions – and interaction – between the subject roles, market institutions, patent system, and experimenter controls.

A separate *documentation* summarizes the software design.

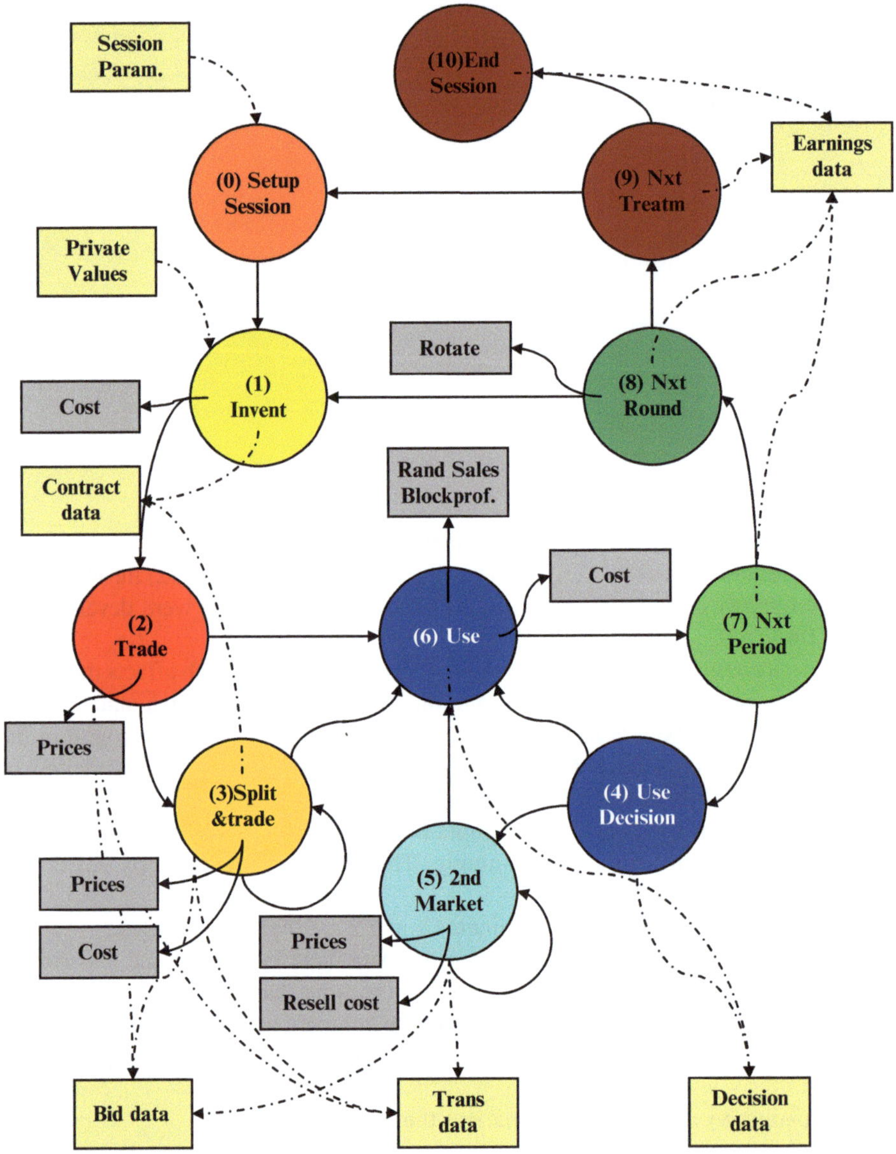

State diagram: Experiment 1

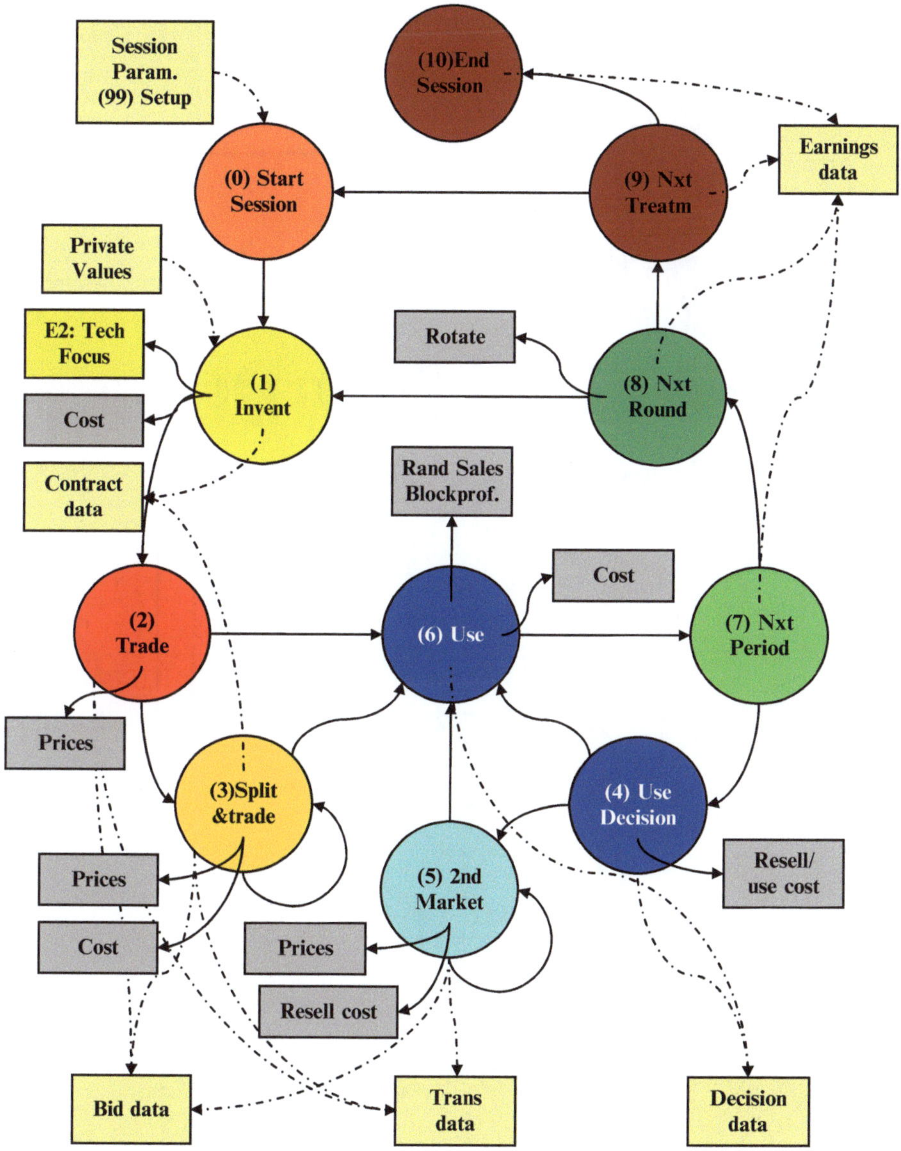

State diagram: Experiment 2

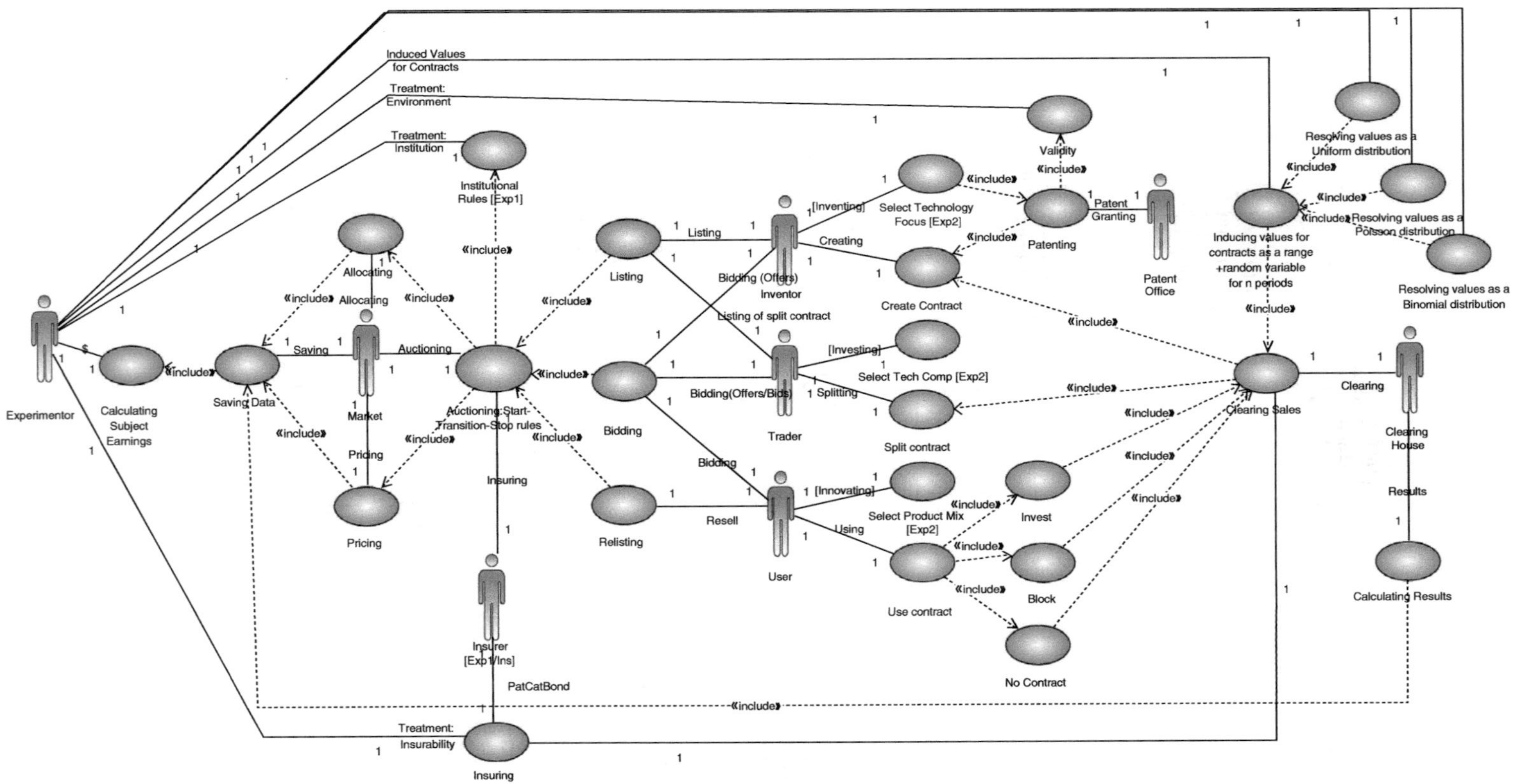

Case diagram: Experiment 1 & 2

Appendix 2

Values

The induced values were calculated with the principle that in each market (each contract) there would be only one agent with the highest value (counted at the midpoint of the value range). The purpose of this was to be able to test the mechanism for allocation to the highest bidder, a criteria for an efficient price system.

For each session, each subject needs high- and low-end values for blocking and investing for standard and high-quality AB and A and B contracts, for three periods and for each contract traded. This means between 30 and 50 values per round and subject. For a 30-round session with 10 subjects, this means more than 10,000 values. Such an environment can only be programmed and all values were calculated in Excel using a special "calculator" to generate appropriate value sets for different sessions. The values calculated using Excel files were then stored as comma-separated text files to be used by the software.

The values are documented in separate files.

Appendix 3

Data

Data was recorded for the whole bidding process, decisions made, and all outcomes of prices, allocations, profits, and value draws.

Complete transaction files were first created using Excel and then imported to Stata X for a database. All statistical analysis was then done in Stata. Most graphs were also done using Stata. Tables were created using Excel and some special figures.

The data is documented in the Stata databases, one for each experiment, and in Excel files as well as raw data files.

Appendix 4

Instructions

The instructions were given first by reading through the enclosed text, and then a lecture/walkthrough was done with the software. Finally, trial rounds were run prior to the start of the experiment.

For both experiments, additional information for each session was provided related to the treatment run. These are here enclosed after the general instruction pages. In Experiment 2, an additional sheet of information was added to explain what choices on technology focus could be done.

During the experiment, the subjects had each a copy of instructions as well as explanation of trading screens and a "flow chart" of the experiment to facilitate what each role could do given the phase of the experiment in each round.

In order to assure that the subjects understood the experiment, a pilot test was performed and only the participants able to make a profit in Roles 1 and 3 were invited to the actual experiment. Initially they had to make a profit in each role but Role 2 – the intermediary trader role – turned out to be very challenging for most (similar to real-world traders). Subjects were informed ahead of time or this training procedure. In the end about 55 students, mostly undergraduate students, participated in the experiment.

Welcome to Today's Experiment!

Once the experiment begins, there will be no talking. This rule is in effect until the experiment ends and you have received your earnings. If you have a question, please raise your hand, and someone will come to assist you.

At this time, please silence all phones and any other electronic devices that may make noise during the experiment. The use of personal electronic devices during the experiment is prohibited.

All prices, costs, and values shown to you during the experiment are in *eDollars*. The more eDollars you earn in the experiment, the more US dollars you will earn. At the end of the experiment, you will be told how your earnings will be converted from eDollars to US$.

The earnings depend on the decisions you make. In addition to your 7$ show-up fee, you will earn 7 $/h if you do not make any losses (lose your capital) plus what you earn by trading in the experiment. If you add up losses greater than the capital given in the experiment, you will lose your show-up fee and the 7 $/h. You may thus end up with 0 US$ from this experiment or, if you make profits, 21$ *plus* trade profits for a 2 h session.

Instructions

In this experiment you will trade (buy and sell) a *contract* in a market.

The contract gives a buyer the right to use a certain "technology" useful for products A and B.

The Contract

The contract is useful to the holder (owner) to *produce* a product A or B for three periods.

The contract is paid for in *two* values: One *fix* amount per period and one *percentage* amount based on the sales of products. (The percentage is here called royalty).

The contract can be used to *Block* another user from producing product A or B. The holder of the contract then receives a profit from the contract that would otherwise be lost to another buyer. This is the "blocking value" of the contract.

Example 1 A contract with price (5, 20%) and used for product sales of 30 costs: 5 each period + 20% of 30 sales = 6 each period, thus 11 each period. This is what the seller will *receive* and what the buyer will *pay* each period.

Example 2 A contract with the same price (5, 20%) and used for blocking sales for another user costs: 5 each period + 20% of 0 sales = 0 each period, thus 5 each period. This is what the seller will *receive* and what the buyer will *pay* each period. Thus, by blocking, the seller will only receive the fix part and the buyer only pay the fix part.

For each contract, you will receive estimates of the values of that contract if you:

1. *Invest* to produce A or B (the use value)
2. *Block* to protect you current sales (the blocking value)

The Value of a Contract

The values are expressed in a *range* from low to high. Values are given for the three periods in the beginning of the round. The actual value that will be used to calculate your revenues (seller) or costs (buyer) is given when the contract is *used* by a holder (Role 3) to Invest or Block and can be any value in the indicated range. The actual value is therefore uncertain at the time of purchase of the contract and resolved only at the end of each period.

The value of investing (use) and blocking ("sit on it") using the contract are thus given as a range [low–high], one range for each type of value for the three periods. These values may differ or be the same over the three periods in given a round. When a contract is used, a random value is given to the contract that is *between* the low and the high estimate for that period. The value has an equal chance (probability) to be anywhere in that range (Table A.1).

The Contract Focus

What is traded are thus contracts on a certain "technology." The technology has the *focus* "AB," meaning that it can be used to produce products of type A *and* B. A contract can also be *split* into two contracts with focus A and B. A new contract AB is created in each round.

The *validity* of a contract means that, if the validity <100%, the technology on which the contract is based may be useless.

The *discount rate* (Disc.Rte) indicates how fast the value of the contract is reduced each period. A rate of 10% means that the value is reduced by 10% each period. This value thus indicated how the value ranges are valued by a given role. To help you estimate the value of a contract, with all its dimensions of uncertainty, validity, and discount rate, there is a "profit calculator" on your screen to estimates your profits given what you are bidding/offering for a contract.

Table A.1 Example of contract value estimates

Control	Focus	Validity (%)	Use	Disc.Rte (%)	Period 1	Period 2	Period 3
1	AB	100	Invest	10	10–19	10–19	10–19
1	AB	100	Block	10	5–10	5–10	5–10

The Roles

The contract sold by Role 1 (one in each round) is of focus "AB." The contracts potentially *split* by Role 2 have focus A and B, respectively. In a similar way, Role 3 can produce *either* product A or product B. This is indicated by the "Product mix" for Role 3. The product mix is the product you can produce in the round. Thus, splitting a contract allows *both* A and B products to be produced, each using a separate contract. If the initial contract is not split, then only A *or* B can be produced, limiting the revenues.

You will be playing *one* of the three roles: 1, 2, or 3. The role will be assigned to you at the beginning of the experiment but it is important to know what all roles are doing since these are the roles you will trade with. There are an unknown number of *rounds* in the experiment. Each round has three *periods*. During a round one, contract is created (Role 1), sold (to Role 2 or 3), and potentially split into two (by Role 2) and the two split contracts sold (to Role 3), and finally the contract(s) are used to produce products "A" and/or "B" for three periods (Role 3). During periods 2 and 3, the contract can be *resold* between Role 3 participants if desired.

The "price" of the contract is given by a *fix fee* and a *percentage* on the revenues from producing A or B, both paid each period to the seller (issuer) by the buyer (holder).

How the Contract Is Traded

The trading is done in an *auction* where buyers place *bids* on a contract and sellers place *offers* ("asks").

For the next section, please look at the screen images of Roles 1, 2, and 3.

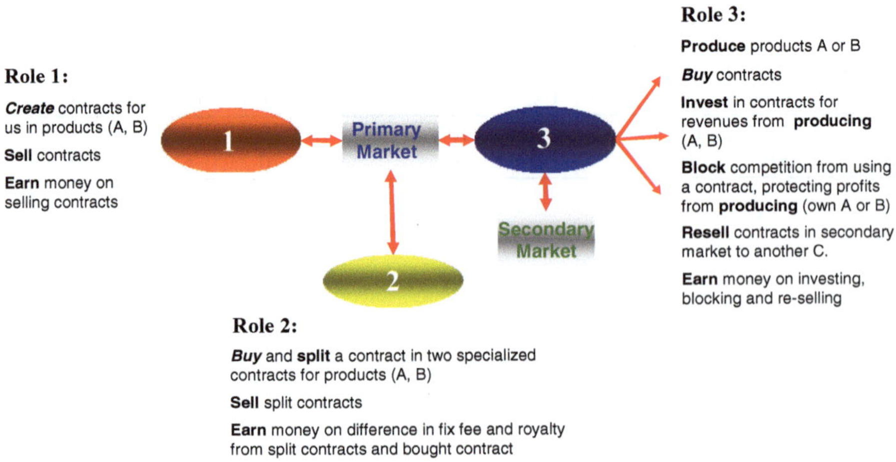

Creating a Contract (Role 1)

There are two steps in creating a contract:

1. Select a quality of the contract ("standard" or "quality"). The cost of creating the contracts are different.
2. Hit the green button to Invest in the contract.

Bidding on Contracts (All)

There are three steps in submitting an offer/bid:

1. Enter you expectations of the value of the contract for each period (1, 2, 3) [in light blue boxes].
2. Enter an offer/bid in two dimensions, fix and percentage ("royalty") [in red/pink boxes].
3. Check your expected profits from your offer/bid [in light green boxes] and submit by clicking the "Send" button. In some cases, there is also an "Accept" and "Reject" button.

Splitting a Contract (Role 2)

Simply split a contract bought by clicking on the "Split" button. Then, sell the two new contracts in the same way as "bidding." Only one contract can be sold at a time.

Using a Contract (Role 3)

At the end of the trading phase, one or two players of Role 3 will own a contract. You can now choose to "Invest" or "Block" or "Resell" a contract. In the first period, you are asked to Invest/Block. If you Invest, then you will receive revenues in the "usage" value ranges of the contract. You will also pay an investment cost and a "production cost" for product A or B and the fix fee and a percentage (royalty) on the revenues for using the contract. If you Block, then you will receive a profit in the "Block" value ranges of the contract. You will only pay the fix fee of the contract since there are no A or B revenues.

Role 3 can also resell the contract to another player of Role 3 type at be beginning of periods 2 and 3. This allows a player to earn money from the revenues of the sale of the contract and the new buyer from Invest/Block value of the contract if that is desirable.

At the end of each period, after Role 3 has used the contract(s) sold in the round, your earnings are calculated and showed. All this information will show-up in the *earnings box*.

Summary

To bid on a contract you do three things:

1. Look at the two value ranges using or blocking with the contract. The actual value will be random within the range and decided when it is used by its holder.
2. Enter *your* expectations for the value (somewhere in the range).
3. Enter you ask/bid, checking the expected profit for using or blocking.

Splitting the Contract – Role 2

When Role 2 has bought a contract in the primary market, then the contract is split into two new contracts, one with focus "A" and another with focus "B." Role 2 now sells the contract, one at a time just like Role 1 has done with the first contract. Role 2 thus is fist a "buyer" and then a "seller" of contracts.

Using the Contract – Role 3

When Role 3 has bought a contract in the primary market, either a contract good for producing both A and B or for either A or B, the contract has to be used. Role 3 now decides whether to "Invest" – which leads to sales from product A or B, depending on which product mix this particular Role 3 has – or to "Block" – which leads to blocking profits but no sales since Role 3 is simply "sitting on" the contract.

After all Role 3 who have contracts (1 or 2) have decided how to use them, then the earnings from the period are calculated for everyone and shown in the earnings box.

These earnings are then added up – or subtracted if negative – from the capital initially given.

If the capital becomes negative, that game is over for that participant, and he/she cannot continue and looses the show-up fee and hourly fee given by keeping the capital.

In the next section, the specific rules of bidding/offering are explained in detail.

Earnings Per Role

The profits for each round are calculated as:

Role 1

+ Revenues from contract = fix + royalty × Sales of A, B this period
– Cost of creating contract = standard/quality contract cost (in period 1)
= Profit

Revenues are the sum of the sales of products A and B. If the contract is standard, then since only one Role 3 can buy a contract you will either have revenues from A or B if it is used to produce. If the contract is quality and Role 2 bought it and split it, you will have sales from A *and* B. However, if Role 3 uses the contract to Block, then the sales are 0 since Role 3 is "now sitting on it." You will then only get the fix fee.

Role 2

+ Revenues from split contract A = fix + royalty × Sales of A this period
+ Revenues from split contract B = fix + royalty × Sales of B this period
– *Cost of contract AB = fix ± royalty × Sales of A ± B this period*
= Profit
– *Validity insurance (not always used)*
= Profit

If Role 3 chooses to Block instead of Invest and use, then only the fix fees are earned from Role 3 and have to be paid to Role 1.

Role 3

This role earns money in three ways.

+ Sales from producing own products (A or B)
– Cost of producing products = fix cost/period + variable × sales (*see calculator*)
= Profit from using the contract
+ Profit from blocking (if blocking otherwise 0)
= Profit from using contract
+ *Profit from reselling (if reselling contract)*
= Profit from contract
– Cost of contract = fix + royalty × sales of products (A or B)
– *Cost of buying contract in secondary market (if done so in period 2 or 3)*
= Profit
– *Validity insurance (not always used)*
= Profit

Session Specific Information

How the Auction Works/DA1

General concept

In an auction, the price of the contract sold is decided by the lowest price a seller is *willing to accept* and the highest price a buyer is *willing to pay*. The sellers price is often called an "offer" or an "ask" and the buyers price most commonly a "bid." Here, we use offer for the seller's price and *bid* for the buyer's price.

Only if the seller and buyer meet with their prices, there is a trade. This can happen in two ways:

1. The seller *reduces* the asking price or the buyer *increases* the bidding price until they meet.
2. The seller *accepts* the buyer's highest bid or the buyer *accepts* the seller's lowest offer.

This Experiment

In the auction in this experiment, it works in a similar way:

1. The seller sends in what they are *willing to accept* based on their *expectations* for each period, of how much the contract could be worth, based on the value range *estimates* given to the seller. The bid has two parts:
   - A *fix* amount asking for each period (does not change with revenues from A or B).
   - A *percentage* of the revenues from A or B, here called *royalty*.
2. The buyers send in what they are *willing to pay* each period for the contract in fix and royalty.
3. A new ask or bid has to be send in every 20 s until the buyer and seller find a price they can agree on. If there is no agreement within 120 s from the first ask sent by the seller, the contract is not sold and the cost of creating the contract has to be paid by the seller.

On your screen, the value of the contract given *your* expectations, discount rate, and tentative bid is calculated and shown in the "profit" box. You must enter expectations for both the use and Block values to send in an ask or a bid. If the profit is positive, you will earn money. If the profit is negative, you will loose money. There is no other way to earn money in this experiment except through profit.

The actual value of the contract is then decided when Role 3 uses the contract (s) to "Invest" or to "Block." This value is shown to everyone in the box "Contract Usage Info."

Example:

Seller's values:

Control	Validity (%)	Use	Period 1	Period 2	Period 3
1AB	100	Invest	5–10	7–15	10–20
1AB	100	Block	2–3	2–3	2–3

Seller selects expectations of: *5, 7, 10* for the three periods and submits a first ask of *5, 20%*.

Buyer selects his or her expectations and submits a bid of *2, 10%*.

Seller accepts *2, 10%.*
Buyer gets the contract and has to pay *2, 10%.*

For treatment DA1:
New rules for buying and selling contracts in the primary market:

1. A seller (*A* or *B*) submits an offer of the contract in fix and royalty%.
2. A buyer (*B* or *C*) submits a bid in the two dimension fix and royalty%.
3. A seller can then accept the buyers bid or the buyer can accept the sellers offer or submit a new, better, offer/bid.

   Rules for new bid:

   (a) A seller must *reduce* the standing (=highest) fix and/or royalty% offer
   (b) A buyer must *increase* the standing (=highest) fix and/or royalty% bid

   The build-in calculator estimates the profit from the trade as before: You enter the expected revenue for the contract in the *blue* boxes and your profit, given your bid/offer (entered in the *red* boxes) show-up in the *green* boxes.

For treatment DA2:
New rules for buying and selling contracts in the primary market:

4. A seller (*A* or *B*) submits an offer of the contract in fix and royalty%.
5. A buyer (*B* or *C*) submits a bid in the two dimension fix and royalty%.
6. A seller can then accept the buyers bid or the buyer can accept the sellers offer or submit a new, better, offer/bid.

   Rules for new bid:

   (a) A seller must *increase* the standing (=highest) fix and/or *reduce* the royalty% offer
   (b) A buyer must *reduce* the standing (=highest) fix and/or *increase* the royalty % bid

   This has the effect that the seller can ask a *minimum* price for the fix (can always be negotiated upwards if the buyer asks a *higher* fix).
   The build-in calculator estimates the profit from the trade as before: You enter the expected revenue for the contract in the *blue* boxes and your profit, given your bid/offer (entered in the *red* boxes) show-up in the *green* boxes.

For treatment PO:
New rules for buying and selling contracts in the primary market:

1. A seller (*A* or *B*) submit a *take-it-or-leave-it offer* of the contract (fix, royalty%). Submitting a contract comes with a cost to the seller each time an offer is submitted.
2. A buyer (*B* or *C*) *accept* or *reject* the take-it-or-leave-it offer in the two dimension fix and royalty%.

3. The first buyer to accept the offer is the winner of the contract. If no-one accepts the offer, the seller will submit a new, lower offer. The offer must decrease in either fix or royalty% or both. Every "round" of offering adds a *cost* to the seller. This goes on until a buyer accepts the offer.

The build-in calculator estimates the profit from the trade as before: You enter the expected revenue for the contract in the *blue* boxes and your profit, given your bid/offer (entered in the *red* boxes) show-up in the *green* boxes.

Rules for Bidding in the Secondary Market (Resell)

Only Role *3* can participate in the resell market.

1. A seller (Role *3*) submits an offer in a *fix price* for the contract owned (that already has a fix and percentage (royalty) price).
2. A buyer (Role *3* that has the same *product mix* (=A or B) as the contract for sale) submits a fix price bid for the contract [with a given fix and percentage (royalty) price].
3. A seller can then accept the buyers bid or the buyer can accept the sellers offer or submit a new, better, offer/bid. Rules for new, better, bid:

   (a) A seller must *reduce* the standing (=highest) fix price
   (b) A buyer must *increase* the standing (=highest) fix price bid

The build-in calculator estimates the profit from the trade. OBS! To estimate profits from Block values, simply enter those expectations in the light blue boxes after you have estimated the use values.

If there are any questions, please raise your hand now.

The experiment will start with a "trial round" to explore the trading screens. That round will not affect your earnings in the experiment.

Comments

Instructions: Complement for Invalidity

28 January, 2009

A contract that has 38% validity means that there is a 38% chance that it will become invalid during a round.

The validity is calculated each period when the contract is used. Since there are three periods this means that there is about a 72.5% chance each period that the contract is invalidated ($0.725^3=0.38$). The same calculation is made if the validity is 93% (0.975^3).

If the contract is invalidated, then this happens sometimes during the period. You can think of a period as a year when Role 3 is producing. The actual time or "date" it is invalidated is a random value from 1 to 365. During the year, Role 3 uses the contract to produce/Block. The year is divided into three "production cycles." All product sales that should have taken place in the period when the contract was invalidated are lost but the cost remains.

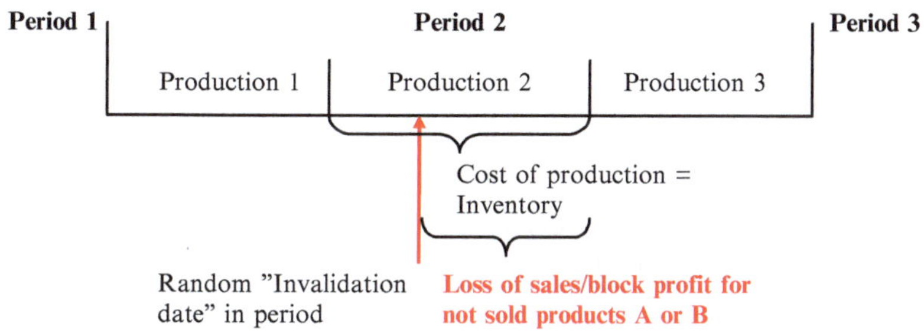

This results in a loss equal to the cost of the remaining inventory. Depending on the invalidation date, this is a maximum of 1/3 of the cost for one *period*.

A Role 3 does not pay any fix or royalty when the contract is invalidated and Role 1 or Role 2 does not earn any money.

When a contract is invalidated, it cannot be resold.

Experiment 2

Changes from Experiment 1 are only in roles and market information.

Roles

Role 1

Chooses a "technology focus" for the contracts traded (and quality/standard). Different focus has different value to a buyer.

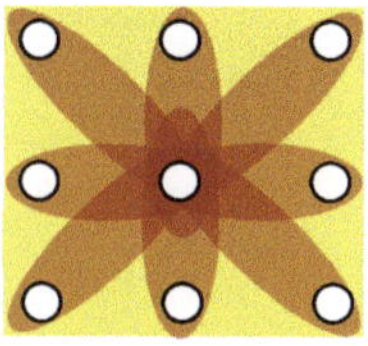

Areas of tech focus

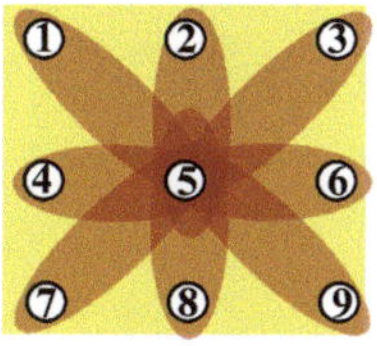

Identification of tech focus: AB1-9

Any of the tech focuses on the "A side" can be "A" and any of the tech focuses on the "B side" can be B. All others have tech focus "AB." The value (expected profits) to a buyer (may) differ for different tech focuses. The "middle" of the tech map indicates the technology a user 3 has today.

The tech focuses are numbered 1–9 for identification purposes in the technology map.

Expectations are set fix at 50%.

Role 2

There are two Role 2 in this experiment.

Expectations are set fix at 50%.

Role 3

No change except

Expectations are set fix at 50%.

Market Info

Current Contract in Primary and Secondary Markets

"Focus" is now "Technology Focus." A "*" is added if the contract is a quality contract.

Transaction List

"Focus" is now "Technology Focus." A "*" is added if the contract is a quality contract.

Secondary market prices are indicated by: Blank if primary market, Price if sold an, "Not Sold" if not sold.

Trading Screens

What you do in the different roles during a round

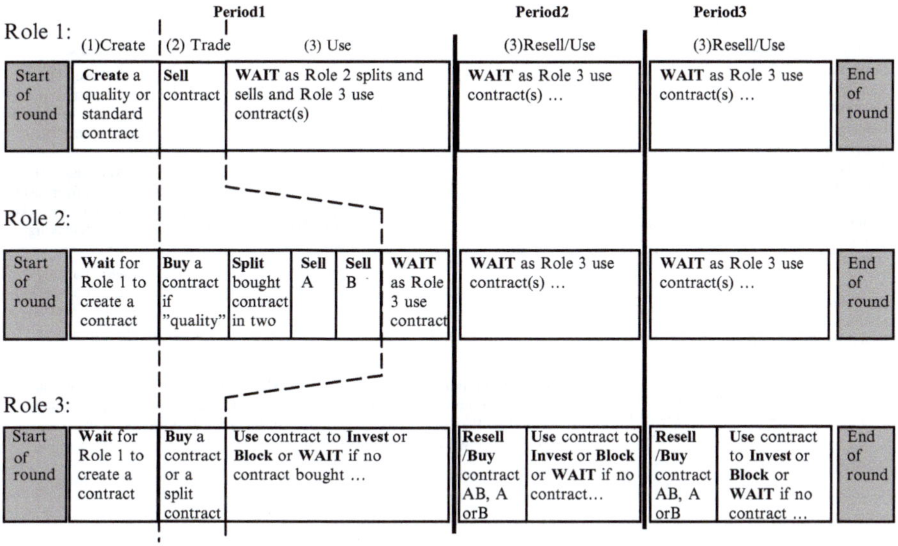

Role: 1

From Personal to Impersonal Exchange

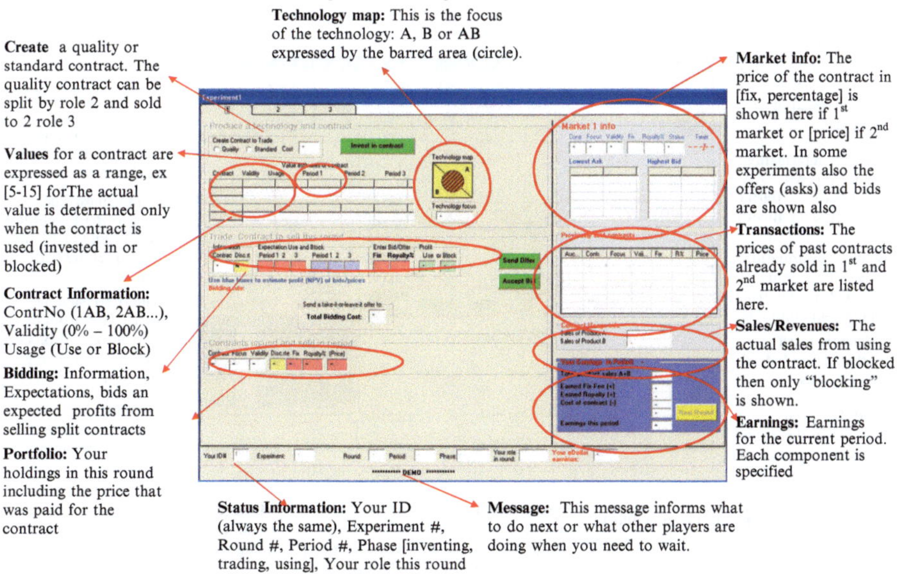

Role: 2

Trade in Ideas

Technology map: This is the focus of the technology: A, B or AB expressed by the barred area (circle).

Values for a contract are expressed as a range, ex. [5-15] forThe actual value is determined only when the contract is used (invested in or blocked)

Contract Information: ContrNo (1AB, 2AB...), Validity (0% – 100%) Usage (Use or Block)

Bidding: Information, Expectations, bids an expected profits from selling split contracts

Estimate: Estimate profits from selling split contracts

Portfolio: Your holdings in this round including the price that was paid for the contract

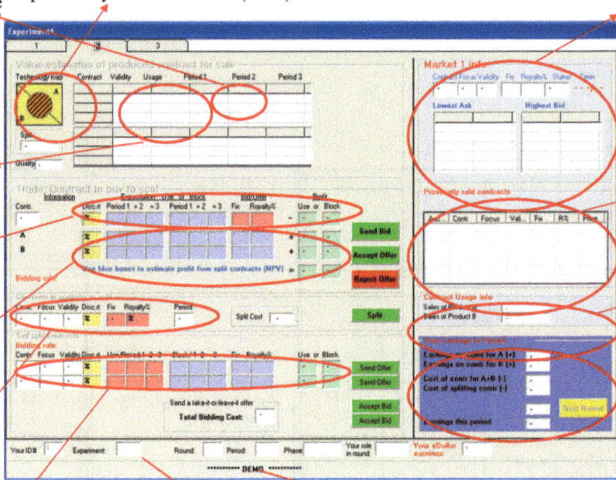

Market info: The price of the contract in [fix, percentage] is shown here if 1st market or [price] if 2nd market. In some experiments also the offers (asks) and bids are shown also

Transactions: The prices of past contracts already sold in 1st and 2nd market are listed here.

Sales/Revenues: The actual sales from using the contract. If blocked then only "blocking" is shown.

Earnings: Earnings for the current period. Each component is specified

Selling split contracts: Information, Expectations, bids an expected profits from selling split contracts

Status Information: Your ID (always the same), Experiment #, Round #, Period #, Phase [inventing, trading, using], Your role this round

Message: This message informs what to do next or what other players are doing when you need to wait.

Role: 3

From Personal to Impersonal Exchange

Technology map: This is the focus of the technology: A, B or AB expressed by the barred area (circle).

Values for a contract are expressed as a range, ex. [5-15] forThe actual value is determined only when the contract is used (invested in or blocked)

Contract Information: ContrNo (1AB, 2AB...), Validity (0% – 100%) Usage (Use or Block)

Bidding: Information, Expectations, bids an expected profits from selling split contracts

Bidding in resell market: Enter bid/ask, check profits

Portfolio: Your holdings in this round including the price that was paid for the contract

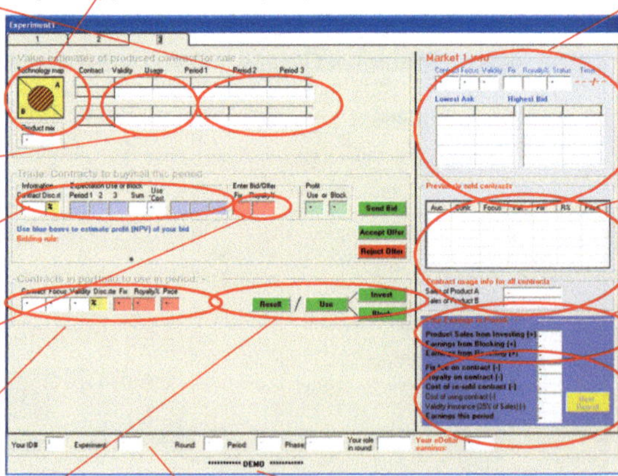

Market info: The price of the contract in [fix, percentage] is shown here if 1st market or [price] if 2nd market. In some experiments also the offers (asks) and bids are shown also

Transactions: The prices of past contracts already sold in 1st and 2nd market are listed here.

Sales/Revenues: The actual sales from using the contract. If blocked then only "blocking" is shown.

Earnings: Earnings for the current period. Each component is specified

Decisions to Use/Resell, Invest or Block: First you can make a decision to use/resell then, if use, to invest/block.

Status Information: Your ID (always the same), Experiment #, Round #, Period #, Phase [inventing, trading, using], Your role this round

Message: This message informs what to do next or what other players are doing when you need to wait.

Bibliography

Arrow, J. K. (1962). Economic Welfare and the Allocation of Resources for Invention. *In NBER The Rate and Direction of Inventive Activity: Economic and Social Factors. Princeton Univ. Press (for the NBER).*

Arrow, K. J. (1963). Uncertainty and the economics of medical care. *American Economic Review, 53*, 941–973.

Arrow, K. J. (1965). *Aspects of the theory of risk-bearing*. Finland: Yrjö Jahnssonin Säätiö.

Arrow, K. J. (1971). Insurance, Risk and Resource Allocation. *Essays in the Theory of Risk-Bearing, Chicago: Markham*, pp. 134–143.

Arrow, K. J. (1986). Rationality of self and others in an economic system. *The Journal of Business, 59*, S385–S399.

Aumann, R. J. & Dreze, J. H. (2009). Assessing strategic risk. *American Economic Journal: Microeconomics, 1*(1), pp. 1–16.

Barzel, Y. (1986). Knight's" Moral Hazard" theory of organization. *Economic Inquiry, 25*, 119–120.

Baumol, W. J. (2002). *The free-market innovation machine: Analyzing the growth miracle of capitalism*. Princeton, NJ: Princeton University Press.

Bernoulli, D. (1738 (1954)). Exposition of a new theory on the measurement of risk. *Econometrica, 22*, 23–36.

Coase, R. H. (1937). The nature of the firm. *Economica, 4*, 386–405.

Coase, R. H. (1946). The marginal cost controversy. London School of Economics and Political Science.

Coase, R. H. (1960). The problem of social cost. *Journal of Law and Economics, 3*, 1–44.

Coppinger, V. M., Smith, V. L., & Titus, J. A. (1980). Incentives and behavior in English, Dutch and sealed-bid auctions. *Economic Inquiry, 18*, 1–22.

Cox, J. C., Roberson, B. & Smith, V. L. (1982). Theory and Behavior of Single Object Auctions. *Research in Experimental Economics, 2*. Greenwich: JAI Press.

Cox, J., Isaac, M. R., Cech, P.-A., & Conn, D. (1994). Moral hazard and adverse selection in procurement contracting. *Games and Economic Behavior, 17*, 147–176.

Crouch, D. D. (2008). The rising size and complexity of the patent document. Research Paper, University of Missouri School of Law.

Davis, D., Holt, C. A., & Villamil, A. P. (1990). Supracompetitive prices and market power in posted-offer experiments. University of Illinois BBER Faculty Working Paper No. 90–1648.

Diamond, P. (1971). A model of price adjustment. *Journal of Economic Theory, 3*, 217–227.

Easly, D., & Ledyard, J. (1979). Simultaneous double oral auction markets. Public Choice Society Meetings, March 1979.

England (1623). English statute of monopolies of 1623, 21 Jac. 1, c. 3, The Original Source of the Anglo-American Patent Law.

Fama, E. F. (1996). Discounting under uncertainty. *The Journal of Business, 69*, 415–428.

Fehr, E., Hart, O., & Zehnder, C. (2009). Contracts, reference points, and competition-behavioral effects of the fundamental transformation. *Journal of the European Economic Association, 7*(2–3), 561–572.

Friedman, M. (1976). *Price theory*. New York: Aldine de Gruyter.

Friedman, D. (1984). On the efficiency of experimental double auction markets. *The American Economic Review, 74*, 60–72.

Friedman, D. (1993). The Double Auction Market Institution: A Survey. In *The Double Auction Market: Institutions, Theories, and Evidence*. Addison Wesley Publishing Company, pp. 3–25.

Friedman, D., & Rust, J. (1993). *The double auction market: Institutions, theories, and evidence*. Reading, MA: Addison-Wesley.

Fudenberg, D., & Maskin, E. (1986). The folk theorem in repeated games with discounting or with incomplete information. *Econometrica, 54*, 533–554.

Gambardella, A., Giuri, P., and Mariani, M. (2005). The value of European patents evidence from a survey of European Inventors. Final Report of the PATVAL EU project, Contract HPV2-CT-2001-00013, January 2005.

Greif, A. (2002). Institutions and Impersonal Exchange. *Journal of Institutional and Theoretical Economics, 158*, 168–204.

Hanson, N. (1971). *Observation and explanation: A guide to philosophy of science*. New York: Harper & Row.

Hanson, N., & Humphreys, W. (1970). *Perception and discovery: An introduction to scientific inquiry*. San Francisco: Freeman, Cooper.

Hart, O. D. (1988). Incomplete contracts and the theory of the firm. *Journal of Law, Economics, & Organization, 4*, 119–139.

Hart, O., & Moore, J. (1988). Incomplete contracts and renegotiation. *Econometrica, 56*, 755–785.

Heckscher, E. (1919). The effect of foreign trade on the distribution of income. *Ekonomisk Tidskrift, 21*, 1–32.

Hurwicz, L. (1960). *Optimality and informational efficiency in resource allocation processes*. Stanford University Press.

Hurwicz, L. (1973). The design of mechanisms for resource allocation. *The American Economic Review, 63*, 1–30.

Kahneman, D., & Tversky, A. (1979). Prospect theory: An analysis of decision under risk. *Econometrica, 47*, 263–292.

Kamien, M. I. (1992). Patent licensing. In *Handbook of Game Theory with Economic Applications*. Elsevier BV, pp. 331–354.

Kamien, M. I., Oren, S. S., & Tauman, Y. (1992). Optimal licensing of cost-reducing innovation. *Journal of Mathematical Economics, 21*, 483–508.

Kamien, M. I., & Schwartz, N. L. (1982). *Market structure and innovation*. Cambridge: Cambridge University Press.

Kamien, M. I., & Tauman, Y. (1986). Fees versus royalties and the private value of a patent. *The Quarterly Journal of Economics, 101*, 471–492.

Kamien, M. I., & Tauman, Y. (2002). Patent licensing: The inside story. *The Manchester School, 70*, 7–15.

Kaplan, A. (1964). *The conduct of inquiry*. San Francisco: Chandler.

Katz, M. L., & Shapiro, C. (1984). How to license a patent. Woodrow Wilson School, Princeton University.

Katz, M. L., & Shapiro, C. (1985). On the licensing of innovations. *The RAND Journal of Economics, 16*, 504–520.

Kaufer, E. (1989). *The economics of the patent system*. Switzerland: Harwood Academic.

Ketcham, J., Smith, V. L., & Williams, A. W. (1984). A comparison of posted-offer and double-auction pricing institutions. *The Review of Economic Studies, 51*, 595–614.

Kharif, O., & Miller, H. (2011). Ericsson Said to Weigh Entering Contest for Nortel Patents. *Bloomberg*. Available at: http://www.bloomberg.com/news/2011-05-17/ericsson-said-to-consider-entering-competition-for-nortel-patent-portfolio.html.

Kingston, W. (2004). *Why harmonisation is a Trojan horse*. London: Sweet & Maxwell/ESC Publishing.

Klee, M. M., & Fairfield, C.T. (1998). Where did the US patent laws come from? *Engineering in Medicine and Biology Magazine, IEEE, 17*(1), pp. 135–139.

Knight, F. H. (1921). *Risk, uncertainty and profit*. New York: AM Kelley.

Krugman, P. (1980). Scale economies, product differentiation, and the pattern of trade. *American Economic Review, 70*, 950–959.

Krugman, P. (1985). *A 'technology gap' model of international trade*. London: Macmillan.

Krugman, P. (1990). Endogenous innovation, international trade and growth. In *Rethinking International Trade*. The MIT Press.

Krugman, P. (1990b). *Rethinking International Trade*. Cambridge, MA: MIT Press.

Laffont, J.-J., & Tirole, J. (1987). Auctioning incentive contracts. *The Journal of Political Economy, 95*, 921–937.

Lamoreaux, N. R., & Sokoloff, K. L. (1999). Inventive Activity and the Market for Technology in the United States, 1840–1920. NBER Working Paper No. 7107.

Lamoreaux, N. R., & Sokoloff, K. L. (2001). Market trade in patents and the rise of a class of specialized inventors in the 19th-century United States. *The American Economic Review, 91(2)*, Papers and Proceedings of the Hundred Thirteenth Annual Meeting of the American Economic Association, 39–44.

Leroy, S. F., & Singell, L. D., Jr. (1987). Knight on risk and uncertainty. *The Journal of Political Economy, 95*, 394–406.

Markowitz, H. (1952). Portfolio selection. *The Journal of Finance, 7*, 77–91.

Maskin, E., & Riley, J. (1984). Optimal auctions with risk averse buyers. *Econometrica, 52*, 1473–1518.

McCabe, K. A., Rassenti, S. J., & Smith, V. L. (1992). Designing a uniform-price double auction: An experimental evaluation. In D. Friedman & J. Rust (Eds.), *The double auction market: Institutions, theories and evidence* (pp. 307–332). Reading, MA: Addison-Wesley.

Merton, R. C. (1973). An intertemporal capital asset pricing model. *Econometrica, 41*, 867–887.

Milgrom, P. R., & Weber, R. J. (1982). A theory of auctions and competitive bidding. *Econometrica, 50*, 1089–1122.

Modigliani, F., & Miller, M. H. (1958). The cost of capital, corporation finance and the theory of investment. *The American Economic Review, 48*, 261–297.

Nordhaus, W. (1969). *Invention, growth and welfare*. Cambridge, MA: MIT Press.

Nordhaus, W. D. (1972). The optimum life of a patent: Reply. *The American Economic Review, 62*, 428–431.

Ohlin, B. (1924). The theory of trade. First published in Swedish as Handelns Teori, Stockholm. pp. 156–168.

Plant, A. (1934). The economic theory concerning patents for inventions. *Economica, 1*, 30–51.

Popper, K. R. (1959). *The logic of scientific discovery*. London: Hutchinson.

Porter, D. et al. (2009). The Design, Testing, and Implementation of Virginia's NOx Allowance Auction. *Journal of Economic Behavior and Organization, 69*(2), pp.190–200.

Radner, R. (1995). Economic survival. Nancy L. Scwartz Lecture, Kellogg School of Management, Northwestern University.

Rassenti, S., Smith, V. L., & Bulfin, R. (1982). A combinatorial mechanism for airport time slot allocation. *Bell Journal of Economics, 13*, 402–417.

Rassenti, J. S., Smith, L. V., & Wilson, J. B. (2002). Demand-side bidding will reduce the level and volatility of electricity prices. *The Independent Review, 6*(3), 441–445.

Reiter, S. (1977). Information and performance in the (new) welfare economics. *The American Economic Review, 67*, 226–234.

Robinson, J. (1977). What are the questions? *Journal of Economic Literature, 15*, 1318–1339.

Razgaitis, R. (2006). U.S./Canadian Licensing In 2005; Survey Results. *LES Nouvelles*.

Razgaitis, R. (2007). U.S./Canadian Licensing In 2006; Survey Results. *LES Nouvelles*.

Samuelson, W. (1986). Bidding for contracts. *Management Science, 32*, 1533–1550.

Samuelson, W. (1987). Auctions with contingent payments: Comment. *The American Economic Review, 77*, 740–745.

Schumpeter, J. A. (1934). The Fundamental Phenomenon of Economic Development. In *The Theory of Economic Development: AN Inquiry into Profits, Capital, Credit, Interest and the*

Business Cycle. University of Illinois at Urbana-Champaign's Academy for Entrepreneurial Leadership Historical Research Reference in Entrepreneurship, pp. 57–94.

Schumpeter, J. A. (1942). *Capitalism, Socialism and Democracy*, New York: Harper.

Schumpeter, J. A. (1994). *History of economic analysis*. London: Routledge.

Sharpe, W. F. (1964). Capital asset prices: A theory of market equilibrium under conditions of risk. *The Journal of Finance, 19*, 425–442.

Smith, V. L. (1962). An experimental study of competitive market behavior. *The Journal of Political Economy, 70*, 322–323.

Smith, V. L. (1968). Optimal insurance coverage. *The Journal of Political Economy, 76*, 68–77.

Smith, V. L. (1976). Bidding and auctioning institutions: Experimental results. In Y. Amihud (Ed.), *Bidding and auctioning for procurement and allocation* (pp. 43–64). New York: New York University.

Smith, V. L. (1981). An Empirical Study of Decentralized Institutions of Monopoly Restraint. *Essays in Contemporary Fields of Economics:* In honor or Emanuel T. Weiler, Eds. Weiler, Horwich, W. Lafayette: Purdue University.

Smith, V. L. (1982). Microeconomic systems as an experimental science. *The American Economic Review*, 72, 923–955.

Smith, V. L. (1994). Economics in the laboratory. *The Journal of Economic Perspectives, 8*, 113–131.

Smith, V. L. (2002). Banquet speech, Nobel dinner. Nobel Foundation.

Smith, V. L. (2003). Markets, Institutions and Experiments. *Encyclopedia of Cognitive Science, L. Nadel (ed.)*, Macmillan Publishers: Nature Publishing Group.

Smith, V. L. (2004). Human nature: An economic perspective. *Daedalus, 133*, 67–76.

Smith, V. L. (2007). *Rationality in Economics: Constructivist and Ecological Forms*, Cambridge: University Press.

Smith, V. L., Suchanek, G. L., & Williams, A. W. (1988). Bubbles, crashes, and endogenous expectations in experimental spot asset markets. *Econometrica: Journal of the Econometric Society, 56*, 1119–1151.

Tirole, J. (1999). Incomplete contracts: Where do we stand? *Econometrica, 67*, 741–781.

Ullberg, E. (2004). Is there a patent beta? Cost of capital evaluation of patent active and patent non-active companies. Working Paper Note.

Ullberg, E. (2009). From personal to impersonal exchange in ideas – Experimental study of trade in organized markets for patents. *KTH TRITA-TEC-PHD 09–006*,180.

Ullberg, E. (2010a). Coordination of inventive and innovative activities through markets with prices: An experimental study of patent markets with transparent prices. CESIS Working Paper Series #231.

Ullberg, E. (2010b). Du passage d'un système d'échanges "personnels" à un système d'échanges "impersonnels" des idées: une étude expérimentale (Encadré 3). In D. Guellec, T. Madiès, and J.-C. Prager (Eds.), *Les marchés des brevets dans l'économie de la connaissance*. Report for the Economic Counsel of the Prime Minister, Paris.

Ullberg, E. (2010b). From personal to impersonal exchange in ideas: An experimental study of patent markets with transparent prices. CESIS Working Paper Series #230.

Ullberg, E. (2010c). The problem of trading patents in organized markets: A dynamic experimental microeconomic system model and informal price theory. CESIS Working Paper Series #229.

Ullberg, E., Rodriguez, E., & Stormby, N. (2002). Risk management – From portfolio strategy to value creating system strategy. *Geneva Papers on Risk and Insurance – Issues and Practice, 27*, 467–476.

Vickrey, W. (1976). Auctions, markets, and optimal allocation. In Y. Amihud (Ed.), *Bidding and auctioning for procurement and allocation*. New York: New York University.

Vickrey, W. (1962). Auctions and bidding games. In *Recent advances in game theory*. Princeton, NJ: Princeton University Press.

WIPO. (1883). The Paris Convention for the protection of industrial property. *WIPO Database of Intellectual Property Legislative Texts*.

WTO. (1994). *The TRIPS Agreement*, Available at: http://www.wto.org/english/tratop_e/trips_e/trips_e.htm.

Author Index

A
Arrow, J.K., 2, 10, 16, 17, 35, 48, 52, 55, 68, 114, 139, 141, 147, 157
Aumann, R.J., 52

B
Barzel, Y., 13, 147
Baumol, W. J., 12
Bernoulli, D., 141, 153

C
Coase, R. H., 4, 7, 12, 18, 19, 35, 43, 45, 66, 67
Coppinger, V.M., 54
Cox, J., 51, 54
Crouch, D.D., 9

D
Davis, D., 54
Diamond, P., 140

E
Easly, D., 54

F
Fama, E.F., 153
Fehr, E., 158
Friedman, D., 13, 54
Friedman, M., 147
Fudenberg, D., 52

G
Gambardella, A., 25, 46

H
Hanson, N., 27
Hart, O.D., 15, 51
Heckscher, E., 10, 119
Holt, C.A., 54
Hurwicz, L., 19, 21, 29, 30, 144

I
Isaac, M.R., 51

K
Kahneman, D., 30, 75, 106, 114, 140, 145, 153, 162
Kamien, M.I., 19, 36, 52
Kaplan, A., 27
Katz, M.L., 19, 49
Kaufer, E., 5, 10, 151, 157
Kingston, W., 155
Knight, F. H., 13, 38, 147
Krugman, P., 2, 3, 10, 35, 111, 119, 120

L
Laffont, J.-J., 51
Lamoreaux, N.R., 11, 40, 65, 66
Ledyard, J., 54

M
Markowitz, H., 52
Maskin, E., 29, 51, 52
McCabe, K.A, 54
Merton, R.C., 52, 153
Milgrom, P.R., 51
Miller, M.H., 3, 52
Modigliani, F., 52
Moore, J., 51

N
Nordhaus, W.D., 2, 9, 10

P
Plant, A., 2, 10, 39
Popper, K.R., 27
Porter, D., 15, 35, 47, 51, 63, 119

R
Radner, R., 51
Rassenti, J.S., 46, 54
Reiter, S., 21
Ricardo, D., 10
Riley, J., 60, 188
Robinson, J., 57, 141, 147

S
Samuelson, W., 51
Schumpeter, J.A., 2, 13, 35, 147
Shapiro, C., 19, 49
Sharpe, W.F., 52
Smith, V., 14, 15, 17, 19–21, 26–31, 35, 36, 39, 41, 47, 52, 54, 55, 59, 63, 75, 79, 119, 120, 134, 141, 144, 147, 157
Sokoloff, K.L., 11, 40, 65, 66

T
Tauman, Y., 19, 52
Tirole, J., 39, 47, 51
Tversky, A., 30, 75, 106, 114, 140, 145, 153, 162

U
Ullberg, E., 13, 35, 63–65, 119–121, 153, 159, 160, 163

V
Vickrey, W., 19, 51

W
Weber, R.J., 51

Subject Index

A

Agent
  characteristics, 50, 66
  innovator, 3, 37, 42, 44, 45, 47, 49–51, 55, 56, 121, 123, 146, 148, 149
  intermediary, 24, 41, 78, 119, 152
  inventor, 46, 47
  investor, 18, 19, 38
  trader, 144, 147
Assets
  choose in action, 7, 67
  choose in possession, 7, 67
  intellectual property rights, 1, 7, 38
  physical assets, 1, 6, 7, 12, 41
Auctions, auction mechanisms
  bankruptcy auction, 19, 66, 159
  combinatorial, 8, 47, 159
  demand-side bidding, 20, 21, 53, 121
  double-auction, 16, 20, 22, 70, 72, 73, 123, 126, 127
  Dutch clock, 22, 53, 73, 74, 127
  optimal auctions, 48, 52, 152
  posted offer, 22, 53
  risk averse buyers, 51
  theory of auctions, 122

B

Bankruptcy
  auction, 19, 66, 159
  limited capital, 24, 38, 73, 75, 100, 128, 151
  rules, 75, 128, 145, 159
Behavior, 3, 9, 21, 25–31, 41, 45, 51, 59, 69, 73, 75, 83, 85, 86, 88, 89, 92, 97, 100, 102, 106, 113, 119–141, 148, 149
Behavioral constraints, 30
Bidding
  adjustment process, 150
  patterns, 58, 81, 93, 109, 115, 120, 134, 138, 140
Bundling, 45, 47, 58, 154

C

Claims, structure, 115, 154
Competition, 6–9, 17, 19, 36–40, 49, 50, 54, 57–59, 64, 67, 68, 70, 76, 85, 88, 98–100, 102, 125, 130, 146, 149
Competitive
  bidding, 37, 55–57, 78, 79, 85, 129
  technology market, 3, 36, 146
Contracts
  incentive, 16, 20, 36, 47, 89, 91, 97, 153
  incomplete, 15, 51, 157
  linear, 15–22, 24, 29, 31, 47, 49, 51, 55, 69, 114, 119, 123, 131, 150
  two-part tariff, 39
Coordination, 2, 3, 7, 14–25, 31, 35, 45, 50, 51, 57–59, 64, 66, 68, 119–123, 129, 131–136, 138–140, 143–145, 148, 149, 156, 160

D

Design parameters, 42, 75–100, 129–138
Disclosure, 1, 5, 8, 9, 24, 37, 41, 44, 45, 65, 121, 122, 134, 140, 145, 149, 154, 157
Dynamic
  efficiency, 14, 64, 69, 107–108
  gains, 2, 24, 35, 36, 57, 63–116, 123, 129, 135, 144, 147–149, 153

Dynamic (*cont.*)
multiple end states, 140
programming, 107
systems, 20, 21, 35, 42, 79, 100, 114, 120, 129, 145

E
Economics
agents, 29, 43–47, 52
bounded rationality, 27
environment, 10, 20, 22–26, 29, 30, 36, 41, 43–50, 57, 69, 103–106, 120, 144, 145
exchange, 1, 7–11, 17, 20, 26–28, 31, 145, 165
institutions, 1
organization, 4, 6, 11–14, 63, 143, 152
principles, 7–11
system design, 30, 35, 37, 41, 58, 64, 65, 79, 119
Emergency power off (EPO), 6, 8, 43, 44, 46
Endogenous, 45, 47, 50, 58, 69–72, 119, 120, 123–125, 129
EPO.*See* Emergency power off (EPO)
Equilibrium, 20, 25, 28–30, 68, 82, 101, 106, 115, 136, 138, 140
Europe, 5, 155, 165
Exchange
gains from exchange, 64, 144
market exchange, 3, 7–8, 13, 51, 63–65, 115, 120, 121, 151
market structure, 52
social exchange, 8–10, 28, 63, 64, 120, 122, 123, 134, 140, 149, 154
Exogenous, 78, 120, 122, 123, 129
Experimental economics
auctions, 20, 159
demand-side bidding, 21, 31, 146

F
Field-of-use, 58
Finance
capital, 48
capital asset pricing model (CAPM), 45, 50, 52, 153
corporate, 153
cost of capital, 13, 50, 57, 153
intertemporal capital asset pricing model (ICAPM), 45, 52, 153
invention and innovation, 13
systemic risk, 13
Firm
theory of the firm, 51
Fixed fee, 3, 4, 9, 16, 18, 21, 25, 40, 42, 47–50, 53–56, 67–69, 72–74, 78–82, 87, 88, 90, 91, 93–98, 100, 107, 113–115, 127, 128, 144, 147–149, 152

G
Geography, 44, 115
Governments, subsidy, 156
Guilds, 4–6, 143, 154

H
Heuristic experiment, 22, 27
Hierarchy, 4, 13, 14, 18, 35, 44, 51, 66, 68, 119, 123, 141, 143, 144, 147

I
Ideas, 1–31, 35, 37, 45, 52, 57, 63–116, 121, 141, 143–144, 147, 152–154, 156, 158–160
Impersonal exchange, 1–31, 35, 37, 39–41, 63–116, 121, 123, 140, 143–144, 147, 149, 152, 160
Incentive, 6, 8–10, 12–16, 20, 21, 23, 24, 29–31, 36–39, 41, 44, 47, 49, 56, 58, 66, 75, 87, 90, 91, 93, 95, 97, 100, 108, 109, 112, 114, 116, 119, 122, 123, 133, 138, 143–145, 147–154
Incomplete contracts, 15, 51, 157
Individual, 10, 46, 49, 65, 69, 75, 87, 121, 123, 129, 133, 134, 153, 157
Induced values
random variation, 87, 90, 91, 94–97
range, 2, 19
Information
presumed validity, 20, 23, 121, 152
Infringement, 9, 11, 48, 154
Innovation
endogenous, 45, 47, 50, 120, 141
exogenous, 120
intentional, 143
Innovator, 3, 37, 41, 42, 44, 45, 47, 49–51, 55–59, 64, 70, 87, 92, 121, 123, 146, 148, 149, 153, 159
Institutional environment, 21, 25, 29, 43, 103–107, 109, 145
Institutional learning, 37–41, 160
Institutions (mechanisms)
DA1, 83, 86, 102, 112, 113, 115
DA2, 95, 113
PO, 102

Insurance, 3, 11, 13, 17, 18, 38, 49, 50, 54, 55, 58, 67, 76, 77, 106, 114, 122, 146, 147, 149, 152, 157
Intellectual property, 1, 4, 6, 7, 13, 38, 63, 65, 121, 147, 157, 165
Intermediary, 24, 41, 78, 119, 152, 153, 163, 172
International patent agreements
  the Paris convention, 6, 41, 43
  WTO/TRIPS, 41, 65
Invention
  growth, 2, 3, 40, 66, 119, 147
  invention, 2, 3, 5, 6, 8–10, 12, 13, 16, 22–25, 31, 35–40, 42–46, 52, 57, 66, 70, 115, 119, 124, 140, 141, 143, 147, 151, 152, 155, 161
  welfare, 4
Inventor, 2, 3, 5, 6, 8–11, 13, 20, 22, 24, 38, 41–51, 56–59, 64–70, 72, 74, 76, 78, 84, 87, 89, 90, 92, 110, 112, 114–116, 121–124, 126, 128–130, 134, 141, 146, 149, 151, 152, 155, 156, 159

J
Japan, 9, 121

K
Knowledge, stock of, 151

L
Legal environment, 10, 25, 29, 36, 41–44, 50, 58, 69–76, 97, 101–106, 115, 120, 123–130, 145
Legal uncertainty, 22–23
License contract, 64
Linear contract, 2–4, 15–22, 24, 29, 31, 36, 37, 39, 47–55, 58, 59, 64, 65, 67–70, 78, 94, 100, 112–115, 119, 121, 123, 124, 129, 131, 144, 146, 147, 150, 152, 153
Loss aversion, 30, 74, 75, 110, 122, 128, 136, 145, 149, 153

M
Market failure
  externalities, external costs and benefits, 4
  price rationing, 12, 14, 18
  subsidies, 151, 156
Markets
  access, 3, 5, 8, 13–14, 17, 18, 20, 24, 38–40, 45, 47, 49, 50, 52, 54, 55, 67, 68, 70, 87, 100, 110–112, 114, 121, 125, 147, 149, 154, 155, 157, 159, 163
  design of mechanisms for resource allocation, 21, 30, 144
  exchange, 3, 7–8, 13, 51, 63–65, 115, 120, 121, 151
  gains, 156, 158
  in ideas, 1, 7, 68, 121, 147, 159
  information efficiency, 20
  optimality, 17, 144
  in patents, 14–25
  resource allocation, 30
  in technology, 7–8, 36–39, 58, 63, 151, 155
Mechanism, 2, 3, 8, 10, 14–31, 35–37, 41–43, 45, 47, 51, 53, 57–59, 63–75, 79, 86, 87, 93, 100, 113–116, 123, 126, 127, 131, 135, 144–150, 152, 154–157, 159, 160, 170
Mechanism design, 3, 14–16, 19–22, 24, 25, 28–30, 35–37, 41, 43, 58, 59, 63–65, 68, 79, 144–146, 148–150
  integration of information and mechanisms, 37, 53, 59, 64, 69, 122
Monopoly, 2, 7, 8, 35, 36, 49, 119, 155
  privilege, 10, 73, 108, 127, 151, 153
Moral hazard, 16, 48, 49, 51, 59, 90, 114

N
Nomothetic experiment, 22, 27, 101

P
Patent
  beta, 45, 50, 55, 153
  documentation, 31, 100, 166
  economics of the patent system, 7–11
  florence, 5
  harmonization, 6, 43, 116, 162, 163
  history, 4–7, 37–39, 66, 121, 145, 150
  ownership, 5, 41, 48, 154
  presumed validity, 6, 20, 22–23, 41–43, 50, 115, 121, 152
  property rights, 1, 4, 7, 8, 13, 38, 65, 66, 75, 143
  right, 2, 3, 13, 17, 19, 23, 24, 35, 44–47, 50, 54, 55, 58, 63, 65, 66, 120, 144, 145
  time, 2, 42, 57, 69, 145, 150
  UK, 5, 40, 155
  value of patents, 25, 39
  Venice, 5, 10, 40, 155
Patent law
  English statue of monopolies 1623, 5, 40, 155
  US amendment 1952, 155
  US patent law of 1832, 65
  Venetian patent law, 1474, 40, 155

Patent licensing
fees, fixed fees, 11, 44, 49, 50, 52, 56–57
licensing of innovations, 38, 44, 65, 115
optimal licensing royalty, 52
Patent licensing firms
Ericsson, 3, 163
Google, 3, 164
Microsoft, 3
Nortel, 3, 159, 163
Patent Offices
EPO, 6, 43
JPO, 43
WIPO, 6, 43
Patent system, 1–14, 20, 22–24, 29, 35–44, 46, 54, 58, 63, 65–66, 94, 110, 111, 115, 116, 120–123, 134, 140, 143–145, 147, 149–158, 160, 166
role of, 156–157
Patent value
uncertain values, 78, 120, 122, 123
uniform distribution, 25, 55, 70, 78
Payment, 15–18, 28, 47, 49, 50, 55, 56, 67, 75, 90, 113, 128, 129
Personal exchange, 1–3, 6, 11, 13, 15, 16, 20, 28, 31, 37, 39, 40, 44, 53, 65–67, 69, 91, 112, 121, 123, 140, 145, 146, 148
Poisson distribution, 126
Policy
developing nations, 154, 156
North-South technology exchange, 160
patent ownership, 154
propositions, 146, 151–158
technology trade policy, 154–155
Political economy, 1, 5, 155–157, 159
Portfolio, 3, 8, 11, 13, 14, 24, 38–40, 46, 47, 49, 50, 52, 66, 70, 74, 112, 113, 124, 127, 128, 159, 163, 164
selection, 52
Presumed validity, 6, 20, 22–23, 41–43, 50, 115, 121, 152
Presumption of validity
high, 43
low, 43
Price
linear prices, 25, 26, 47, 56, 58, 82–84, 100–106, 108, 113–114, 133, 140, 148, 149
price theory, 3, 15, 18, 31, 36, 37, 51, 54, 64, 113–114, 139, 146–147, 149, 150, 157
transparent, public, 1, 2, 12–14, 22, 25, 29, 35, 36, 41, 57, 63, 64, 108, 114, 119–122, 146
Price system
cost of the price system, 57, 173
transaction costs, 4
Privileges
secrecy, 151
selection of, 72–73
subside, 151
Product, 1, 35, 63, 119, 144, 151
Productivity, 4–7, 151
Prospect theory, 30, 38, 75, 106, 114, 122, 136, 140, 145, 149, 162
Publication, 1, 9, 46, 63, 151, 156

R

Results, 2, 3, 21, 36, 46, 64, 65, 72, 73, 79, 83, 85–88, 90, 92–94, 96, 100–112, 114, 120–123, 126, 133, 136, 138–140, 143–149, 152, 153, 155, 182
Risk
discounting, 52
market access risk, 13, 24, 39, 47, 49, 54, 67, 100, 111, 112, 114
mitigation, 50, 58
profit, 13, 14, 72, 82, 83, 85–88, 94–95, 102, 153
risk and uncertainty, 12–14, 17, 18
systemic risk, 12–14, 45, 58, 63, 102, 112, 120, 155
Risk bearing
risk bearing, 14, 16, 18, 22, 39, 45, 57, 64, 68, 110, 114, 116, 144, 146, 152
risk sharing, 3, 15–18, 22, 24, 39, 47–49, 52, 56, 57, 67, 82, 93–94, 110, 114–116, 144, 152
Royalty, 3, 37, 67, 124, 144, 152
Rules, 4, 12, 14, 20, 21, 28, 30, 31, 37, 41, 47, 52, 53, 58, 59, 64, 68, 69, 73–75, 93, 108, 115, 121–123, 127–129, 138, 144–146, 148, 152, 159, 177, 180, 184

S

Search space, 25, 72, 119–121, 123, 129, 133
search area, 2
Service, 1, 2, 4, 7, 12, 13, 17, 24, 46, 63, 65, 112, 116, 156
Social exchange, 8–10, 28, 63, 64, 120, 122. 123, 134, 140, 149, 154
Social gains, 36, 44, 66, 87, 144, 147
Software
design, 166
environment, 28, 69, 78, 79, 120, 123, 170

Specialization, 1, 10–14, 45, 120, 121, 147, 152, 156
Structural change, 9, 36, 63, 115, 153, 154
Subjects
  home work, 78
  payment, 74–75, 128–129
  selection, 7
  training, 29, 79, 129, 172

T
Tax, 40, 153
Technology
  area, focus, 120–124, 130–133, 135, 136, 138–140, 149
  change, 5, 11–13, 36, 39, 40, 45, 50, 57, 58, 64, 65, 72, 87, 106, 111, 116, 123, 130, 133, 149, 152, 155
  gap, 111
  growth, 3, 40, 66, 116, 119, 147, 156, 159
  human command over nature, 19, 25, 69, 144, 154, 160, 165
  human use of forces of nature, 57
  space, search space, 24–25, 120, 138
  trade policy, 154
Trade
  and patents, 12, 22, 152
  system, 3, 10, 11, 58, 64–66, 147, 149, 152, 156, 157
  and technology, 41, 66, 151, 152
Trade in ideas
  developing economies, 7
  North, 5, 154, 159
  South, 154, 159
Trader, 15, 21–24, 37, 38, 40–42, 45, 47, 49, 50, 55–57, 59, 66, 69, 70, 74, 76, 78, 82, 84, 87, 92, 93, 107, 111, 112, 114, 116, 121, 123, 124, 126,128–130, 146, 147, 149, 152, 153, 162, 163 172
Transaction cost, 4, 11, 76, 86, 99, 130
Transferrable rights, 9
TRIPS, 1, 6, 7, 36, 41, 65

U
Uncertainty, 10, 12–14, 18, 19, 21–23, 30, 35, 38, 42, 43, 45, 51, 54, 55, 58, 63, 67,86, 88, 94, 102, 106, 112, 114, 122, 144, 147, 155, 174
USA, 5, 6, 8, 9, 11–13, 16, 19, 23, 40, 41, 43, 44, 48, 63, 64, 66, 110, 121, 155, 160, 165

V
Validation, state, 46
Venice, 5, 7, 10, 40, 155

W
Willingness
  to accept, 22, 37, 53, 69
  to pay, 21, 22, 37, 53, 55, 69, 101, 121
  to search, 3, 119, 122, 133, 134, 149, 153
World trade organization (WTO), 1, 6, 36, 41, 65
WTO.*See* World trade organization (WTO)

MIX
Papier aus verantwortungsvollen Quellen
Paper from responsible sources
FSC® C105338

If you have any concerns about our products,
you can contact us on
ProductSafety@springernature.com

In case Publisher is established outside the EU,
the EU authorized representative is:
Springer Nature Customer Service Center GmbH
Europaplatz 3, 69115 Heidelberg, Germany

Printed by Libri Plureos GmbH
in Hamburg, Germany